TOTALITY

Totality

Eclipses of the Sun

Second Edition

Mark Littmann
Ken Willcox
Fred Espenak

New York   Oxford
OXFORD UNIVERSITY PRESS
1999

Oxford University Press

Oxford   New York
Athens   Auckland   Bangkok   Bogotá   Buenos Aires
Calcutta   Cape Town   Chennai   Dar es Salaam   Delhi
Florence   Hong Kong   Istanbul   Karachi   Kuala Lumpur
Madrid   Melbourne   Mexico City   Mumbai   Nairobi   Paris
São Paulo   Singapore   Taipei   Tokyo   Toronto   Warsaw

and associated companies in
Berlin   Ibadan

Copyright © 1999 by Mark Littmann,
Ken Willcox, and Fred Espenak

Published by Oxford University Press, Inc.,
198 Madison Avenue, New York, New York 10016

Oxford is a registered trademark of Oxford University Press, Inc.

Library of Congress Cataloging-in-Publication Data
Littmann, Mark, 1939-
    Totality : eclipses of the sun / by Mark Littmann and Ken
Willcox, Fred Espenak. -- 2nd ed.
        p.   cm.
    Includes bibliographical references and index.
    ISBN 0-19-513178-9 (cloth)
    ISBN 0-19-513179-7 (pbk.)
    1. Solar eclipses. I. Willcox, Ken, 1943- II. Espenak, Fred. III.
Title.
    QB541 .L69   1999
    523.7'8--dc21   98-53239

1 3 5 7 9 8 6 4 2

Printed in the United States of America
on acid-free paper

Ken Willcox, just 55 years old, passed away only a few months before this book was published. He died a year to the day after the Caribbean eclipse of February 26, 1998, the last of the five major eclipse expeditions he led, always with his special enthusiasm, care, and warmth.

We treasure the memories of our voyages and work together, Ken.

From Mark Littmann
In memory of my mother, Muriel Stein Littmann, and my father, Lewis Littmann, M.D.

From Ken Willcox
To my wife Sara, my daughter Kenna Turner, and my son Jon Willcox

From Fred Espenak
To my father, Fred Espenak Sr., who has always encouraged me to follow my own path. And in memory of my mother, Asie J. Espenak, and my sister, Nancy J. Davies. I miss you both.

Contents

Foreword

What adjective, what superlative hasn't been badly overworked in attempts to describe total eclipses of the Sun? Not only are these eclipses utterly foreign to humdrum life-experiences, and thus confound story-telling, their unworldly coloring of our atmosphere creates equally strange preludes and postludes, which also defy description. So I live in perpetual purgatory because of my inability to paint a succinct word portrait, at least to my satisfaction, of the total-eclipse experience. After having seen more than a dozen eclipses—from land, sea, and air—words still fail me.

Yet, why can I read a resume of Botticelli's "Venus Rising from the Sea" and appreciate, at least at an elementary level, the critic's sense of that painting? Or why can I read a review of new music and come away with an idea of what it sounds like and how I might react to it?

Maybe we journalists and poets haven't seen enough eclipses to formulate a vocabulary and a syntax. (I believe Eskimos have some two score words to precisely characterize ice in its various forms. We do not have 40 different words to precisely characterize eclipses!)

Thus the authors of *Totality* should take a bow. Their book does the best job yet of capturing all eclipse experiences while making the mechanics of the phenomenon understandable to everyone. Its first edition came out in 1991 and immediately became one of my standard references—those in my office that I can grab while holding onto the telephone. I don't think *Totality* has ever failed me when I've had some anxious questioner on the line or when I've needed a quick fact-check.

So it delights me that Mark Littmann and Ken Willcox decided to produce an updated, enlarged edition. And it delights me doubly that they've taken on Fred Espenak as a third author. He has set a new standard for eclipse prediction and map-making; his charts showing where eclipses will occur are both precise and *really* useful!

A consciousness of eclipses has been around as long as sentient beings, albeit a malevolent consciousness for perhaps 99.9 percent of our existence on Earth. Only for the past 3,000 years or so have eclipses been understood as a predictable, repetitive, natural phenomenon. For some cultures, this recognition of inevitability and rhythm washed away fears. For others it didn't; eclipses are still anticipated with foreboding and accompanied by taboos among many of our world's peoples.

How incredibly macabre it must be for those societies to see tourists flocking to their country for the opportunity to stand in the dreaded shadow of the Moon. Sociologically, this is a new phenomenon indeed. Such eclipse migrations are little more than a generation old, a thousandth of one percent of human history. Dating the beginning is precise: the voyage of the cruise ship *Olympia* in 1972, the first eclipse mass-transit system.

People can now regularly and conveniently travel not only to where an eclipse will occur, but to where it is likely to occur in a clear sky. So the opportunity to view nature's greatest spectacle has become as predictable as the eclipse itself! You simply buy a ticket and go to it as if it were a venue on any ordinary holiday. (In February 1998 I estimated that some 20,000 folks boarded cruise ships to witness an eclipse in the Caribbean, and who knows how many thousands of others opted for land packages.)

No one should pass through life without seeing a total solar eclipse. It's not a "science thing;" it's an experience as profound as recognizing a bird's voice or appreciating a contrary point of view. It's simply something that should be done.

Please don't believe that you've seen an eclipse because of a picture on TV or in a magazine. All images, except those live and in your mind's eye, are parodies of the real thing. I can't stress that thought too strongly. You've got to witness this cosmic spectacle for yourself; no intermediaries are allowed!

Each total solar eclipse is unique, and each one I've seen is indelibly etched in my mind. But equally so is the place, the people, and the ambience of the moment—from maneuvering a ship in the Celebes Sea or an aircraft over Finland to setting up a camera next to a dung heap on Java.

With *Totality* you are excellently prepared to begin, or continue, your quest for darkness after daybreak. **Warning:** This pursuit is as addictive and insatiable as eating that first potato chip.

Leif J. Robinson
Editor in Chief, *Sky & Telescope*

Acknowledgments
for Second Edition

To the readers of the first edition of *Totality: Eclipses of the Sun*: thanks for your strong interest and kind words. It is because of you that we have this opportunity to update and expand our book in a new edition. We hope you like it too.

Expert advice for all aspects of this new edition came graciously and abundantly from Charles Lindsey (again!), Solar Physics Research Corporation; and from Joe Hollweg, University of New Hampshire.

Other distinguished solar physicists generously helped us with special sections of the book: Jay Pasachoff, Alan Clark, Jeff Kuhn, Jack Zirker, Gary Rottman, Don Hassler, Richard Fisher, Larry November, Serge Koutchmy, David Dunham, and Karen Harvey.

Some talented astrophotographers were exceptionally generous in allowing us to use their remarkable work: Steve Albers, Randy Attwood, Michael F. Barrett, Ken Bertin, Serge Brunier, Dennis di Cicco, Stephen J. Edberg, Gary Emerson, William Emery, Jacques Guertin, Johnny Horne, Timothy D. Kelley, Jack Newton, Jay M. Pasachoff, Ernie Piini, and Charles Simpson. Steele Hill of NASA supplied pictures from the SOHO mission.

Solar physicist Charles Lindsey used computer graphics to design most of the fine diagrams for the first edition. With those diagrams as models, Tom Wallin and Will Fontanez of the University of Tennessee Cartographic Services Laboratory skillfully and creatively used computer graphics to execute the illustrations for this new edition. Special artwork was created by Sheri Flournoy and Josie Herr.

We are grateful to the staff at Oxford University Press for their trust in this project and for their skill: Joyce Berry, senior editor; Kim Torre-Tasso, production editor; Peter Grennan, copy editor; Susan

Lin, editorial assistant; and Michele LaForge, editor, who set this project on its way. Alexa Selph, of Atlanta, provided the excellent index. Our thanks to Charles Hamilton, director of the University of Hawaii Press, publisher of our first edition, for facilitating our opportunity for a second edition.

Please refer to the acknowledgments for our first edition to more fully appreciate the extraordinary help we received as this book was born.

—*ML, KW, and FE*

I am deeply grateful to my wife Peggy for her careful reading of the manuscript and for her wise advice.

Peggy, Beth, and Owen: sorry about those many late nights and working weekends. Thanks for your loving support.

To my colleagues in the School of Journalism and the College of Communications at the University of Tennessee, thanks for your encouragement in the course of this project. A special note of gratitude to Jim Crook, Director of the School of Journalism; Dwight Teeter, Dean of the College of Communications; Paul Ashdown, Professor of Journalism; and Lee Riedinger, Head of the Department of Physics and Astronomy, who did so much to make this project feasible.

The University of Tennessee Office of Research graciously provided a grant to help with expenses for this new edition of *Totality*.

To Ken Willcox and Fred Espenak, my partners on this project, thanks for your great knowledge, hard work, and gracious spirits during difficult times.

—*ML*

My wife, Sara, as she does with everything in my life, has raised my confidence and inspired me to do my best. The stability she adds to my life allows me to cope with every situation.

Special thanks to Dr. Eli Maor, professor of mathematics, Loyola University, Chicago, for his persistence in urging a second edition of *Totality*.

With unconditional love through a decade, Dr. and Mrs. Malcolm Granberry of Houston have opened their home and shared their lives with Sara and me as I have been undergoing treatments for lymphoma at M. D. Anderson Cancer Clinic. I now know that as important as the medicine that doctors give you is, even more important are the prayers and love of friends and family in helping you to get well.

I would like especially to thank my employer, Phillips Petroleum Company, and all of my supervisors who have stood by me and even provided an Angel Flight to Houston when conditions warranted it.

And thanks to Berry Beaman, past president of the Astronomical League, for his support of the eclipse trips I arranged so that you could view God's most spectacular celestial event.

—KW

Acknowledgments
for First Edition

This book—perhaps no book—is the work of the authors alone. There are very special people who, by their wealth of knowledge, their wisdom, their creativity, and their graciousness made the completion of our book possible. To these people, there is no truly adequate way to express appreciation. We can only try to single out some of them so that you too can know who helped us so greatly.

Ruth S. Freitag of the Library of Congress was absolutely indispensable. Extraordinarily knowledgeable in the history of astronomy, she supplied references, identified little-known scientists of the past, found pictures, submitted to an interview on her eclipse experiences, and then reviewed the entire manuscript.

A remarkable range of vital contributions came from Charles A. Lindsey, a distinguished solar physicist at the Institute for Astronomy, University of Hawaii. Not only did he review the manuscript, he also created most of the book's diagrams (rendered camera ready by Wendy Nakano) and patiently endured interviews on the use of eclipses in his solar research.

Dennis di Cicco of *Sky & Telescope* magazine reviewed a draft of the book and also contributed some of his superb photography. We profited enormously from his, Charlie's, and Ruth's expertise and wise advice.

In our research, we consulted experts in a wide variety of fields. Some generously contributed vignettes to our book, giving it a richness it would not otherwise have. We are deeply grateful to Lucian V. Del Priore, M.D., Stephen J. Edberg, Alan D. Fiala, Carl Littmann, Laurence Marschall, and Jay M. Pasachoff.

A special note of thanks to John R. Beattie. As he recounted his observing experiences, he provided such a ringing moment-by-moment description of a total eclipse that it became the nucleus of our first chapter.

For this book, we interviewed some of the most experienced eclipse veterans from the ranks of both professional and amateur astronomers. Their valuable perspectives form the heart of the chapters on observing a total eclipse, modern scientific uses for solar eclipses, and eclipse photography. Often they kindly contributed their eclipse photographs for use in this book and reviewed these chapters to reduce our errors. Their names are mentioned in those chapters, but they deserve additional recognition and gratitude here: Jay Anderson, Eric Becklin, Richard Berry, David W. Dunham, Richard Fisher, William C. Livingston, George Lovi, Frank Orrall, Leif J. Robinson, Virginia and Walter Roth, Gary Rottman, and Jack B. Zirker. Each of these experts suggested others who have contributed significantly to public appreciation of eclipses or to solar research. We regret that we did not have time to interview all the veterans they suggested.

There are other eclipse veterans and specialists who kindly helped us with information and ideas: Anthony F. Aveni, Kenneth Brecher, Andrew P. Fraknoi, Kevin Krisciunas, Robert S. Harrington, Sabatino Sofia, and E. Myles Standish, Jr.

Two science writers graciously reviewed the whole manuscript to help us improve our presentation. Special thanks to Beatrice C. Owens and Ann L. Rappoport.

Without great libraries and library services, research for books such as this is impossible. Enormous gratitude to the Loyola/Notre Dame Library in Baltimore. A more talented and gracious staff is hard to imagine. We wish we could list all the librarians who helped so significantly with this project. We are also grateful to the U.S. Naval Observatory Library for allowing so many of its books to migrate to us for this project. Thanks, too, to Gerry Grimm of Baltimore and to John Carper of the John G. Wolbach Library, Harvard College Observatory, for their contributions.

Eclipse lore crosses all boundaries, and we needed help from careful and fast-working translators. Thanks to David and Esther Littmann of Detroit and to Geoffrey K. Gay and Richard E. McCarron of Loyola College.

Of course, a book does not appear without the faith, encouragement, and insight of a publisher and an editor. Our editor was Iris Wiley, executive editor of the University of Hawaii Press. With patience, wisdom, and some splendid editing, she guided this project at every stage. We are deeply grateful to her and to the skillful team at the University of Hawaii Press.

—ML and KW

I am profoundly grateful to Peggy, my wife, without whose love, self-sacrifice, insight, and encouragement, this project would have been impossible. To Beth and Owen, thanks for their interest and understanding and love throughout this project. I would also like to thank Jane Littmann, Tom Owens, and Paul Rappoport for their help through the writing of this book.

To Ken Willcox, my admiration for his courage, dedication, and talent. It is a pleasure and honor to work with him.

This book was written while I was teaching astronomy at Loyola College in Baltimore. I am very grateful to Professor Helene Perry, former chairman of the physics department, and to all the members of the physics faculty and staff for their encouragement and friendship.

In the midst of this project, I received a remarkable offer from the School of Journalism at the University of Tennessee. To Professors George Everett, James Crook, and my new colleagues in the College of Communications, I am deeply honored that you have invited me to join you.

—ML

Without the support, encouragement, and understanding love of my wife Sara, my participation in this project would not have been possible. She is a very big part of what I am. I am profoundly grateful to my father, Delbert Willcox, who, when I was young, taught me much of what I know about photography; to my mother, Madelyn Burgess Willcox, who encouraged my interest in science and astronomy; and to my late aunt, Helen Burgess, who gave me a three-inch Moonscope for Christmas in 1955 that stimulated my curiosity about God's universe.

Special thanks to Phillips Petroleum Company, my employer, for its dedication to improving science education in America, and to the Astronomical League for the wealth of opportunities I have received as a member and for the support I have received during my terms as vice president and president.

This book was written during a most difficult time in my life: throughout its writing I was undergoing treatment for cancer. The attention I was able to give the book is a direct result of unselfish love provided by Dr. and Mrs. Malcolm Granberry of Houston, who hold a special place in the lives of my family. The most credit for this book, and for my participation in it, belongs to Mark Littmann, whose gracious and professional guidance provided a challenge that helped lift me above my fears of what the future might hold.

Through it all has come a fuller realization that people are the most important creation in the universe, and that my Lord has put me in contact with some of His best.

—*KW*

TOTALITY

1

The Experience of Totality

Some people see a partial eclipse and wonder why others talk so much about a total eclipse. Seeing a partial eclipse and saying that you have seen an eclipse is like standing outside an opera house and saying that you have seen the opera; in both cases, you have missed the main event.

—Jay M. Pasachoff (1983)

First contact. A tiny nick appears on the western side of the Sun.[1] The eye detects no difference in the amount of sunlight. Nothing but that nick portends anything out of the ordinary. But as the nick becomes a gouge in the face of the Sun, a sense of anticipation begins. This will be no ordinary day.

Still, things proceed leisurely for the first half hour or so, until the Sun is more than half covered. Now, gradually at first, then faster and faster, extraordinary things begin to happen. The sky is still bright, but the blue is a little duller. On the ground around you the light is beginning to diminish. Over the next 10 to 15 minutes, the landscape takes on a steel-gray metallic cast.

As the minutes pass, the pace quickens. With about a quarter hour left until totality, the western sky is now darker than the east, regardless of where the Sun is in the sky. The shadow of the Moon is approaching. Even if you have never seen a total eclipse of the Sun before, you know that something amazing is going to happen, is happening now—and that it is beyond normal human experience.

Less than 15 minutes until totality. The Sun, a narrowing crescent, is still fiercely bright, but the blueness of the sky has deepened into blue-gray or violet. The darkness of the sky begins to close in around the Sun. The Sun does not fill the heavens with brightness anymore.

Five minutes to totality. The darkness in the west is very notice-

Partial phases of the November 3, 1994, total solar eclipse from central Bolivia. [3.5-inch Questar and Thousand Oaks Type 3 solar filter, 1260 mm focal length, f/14, 1/250 second, with ISO 400 film. © 1994 Ken Willcox]

able and gathering strength, a dark amorphous form rising upward and spreading out along the western horizon. It builds like a massive storm, but in utter silence, with no rumble of distant thunder. And now the darkness begins to float up above the horizon, revealing a yellow or orange twilight beneath. You are already seeing through the Moon's narrow shadow to the resurgent sunlight beyond.

The acceleration of events intensifies. The crescent Sun is now a blazing white sliver, like a welder's torch. The darkening sky continues to close in around the Sun, faster, engulfing it.

Minutes have become seconds. The ends of the bare sliver of the Sun break into individual dots of intense white light—Baily's Beads—the last rays of sunlight passing through the deepest lunar valleys. Opposite the beads, a ghostly round silhouette looms into view. It is the dark limb of the Moon, framed by a white opalescent glow that creates a halo around the darkened Sun. The corona, the most striking and unexpected of all the features of a total eclipse, is emerging.

Along the shrinking sliver of the Sun, the beads flicker, each lasting but an instant and vanishing as new ones form. And now there is only one, set like a single diamond in a ring. The remaining one small dot of sunlight fades as if it were sucked into an abyss.

Totality.

Where the Sun once stood, there is a black disk in the sky, outlined by the soft pearly white glow of the corona, about the brightness of a full moon. Small but vibrant reddish features stand at the eastern rim of the Moon's disk, contrasting vividly with the white of the corona and the black where the Sun is hidden. These are the prominences, giant clouds of hot gas in the Sun's lower atmosphere. They are always a surprise, each unique in shape and size, different yesterday and tomorrow from what they are at this special moment.

Baily's Beads at the February 26, 1998, total eclipse, aboard the *MS Statendam* off Curaçao. [600 mm F/4 Nikon with TC-301 2x converter, 1200 mm focal length, f/11, 1/250 second, with ISO 400 film. © 1998 Johnny Horne]

Diamond Ring Effect at the February 26, 1998 total eclipse from Oranjestad, Aruba. [Unitron 60 mm f/15 refractor, 900 mm focal length, f/15, 1/60 second, with ISO 64 film. © 1998 Michael F. Barrett]

Corona of the Sun during the November 22, 1984, total eclipse from Papua New Guinea. [3.5-inch Questar, 1260 mm focal length, f/14, 1/2 second, with ISO 200 film. © 1984 Randy Attwood]

You are standing in the shadow of the Moon.

It is dark enough to see Venus and Mercury and whichever of the brightest planets and stars happen to be close to the Sun's position and above the horizon. But it is not the dark of night. Looking across the landscape at the horizon in all directions, you see beyond the shadow to where the eclipse is not total, an eerie twilight of orange and yellow. From this light beyond the darkness which envelops you comes an inexorable sense that time is limited.

Now, at the midpoint in totality, the corona stands out most clearly, its shape and extent never quite the same from one eclipse to another. And only the eye can do the corona justice, its special pattern of faint wisps and spikes on this day never seen before and never to be seen again.

Yet around you at the horizon is a warning that totality is drawing to an end. The west is brightening while in the east the darkness is deepening and descending toward the horizon. Above you, prominences appear at the western edge of the Moon. The edge brightens.

Suddenly totality is over. A dot of sunlight appears. Quickly this heavenly diamond broadens into a band of several jewels and then a sliver of the crescent Sun once more. The dark shadow of the Moon silently slips past you and rushes off toward the east.

It is then you ask, "When is the next one?"[2]

2

The Great Celestial Cover-Up

If God had consulted me before embarking upon creation, I would have recommended something simpler.

—Alfonso X, King of Castile (1252)

A total eclipse of the Sun is exciting and even profoundly moving.

But what causes a total solar eclipse? The Moon blocks the Sun from view. And that is all you absolutely need to know to enjoy a solar eclipse. So you can now skip to the next chapter.

If however you are reading this paragraph, you are right: there is more to tell—about dark shadows and oblong orbits and tilts and danger zones and amazing coincidences. Yet before you venture further, promise yourself one thing. If for any reason your eyes begin to glaze over, you will stop reading this chapter immediately and go right on to the next. You must not let celestial mechanics, or this explanation of it, stand in the way of your enjoyment of the wild, wacky, and wonderful things people have thought and done about solar eclipses.

Moon Plucking

How big is the Moon in the sky? What is its angular size?

Extend your arm upward and as far from your body as possible. Using your index finger and thumb, imagine that you are trying to pluck the Moon out of the sky ever so carefully, squeezing down until you are just barely touching the top and bottom of the Moon, trapping it between your fingers. How big is it? The size of a grape? A peach? An orange?

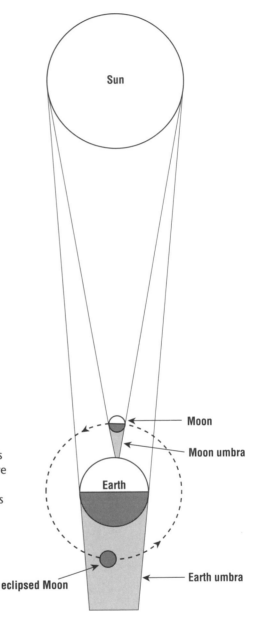

A total solar eclipse occurs when the Moon's umbra touches the Earth. A lunar eclipse occurs when the Moon passes into the Earth's shadow (umbra). The relative sizes and distances of the Sun, Moon, and Earth in this diagram are not to scale.

It is the size of a pea. (You can win bets at cocktail parties with this question.) The Moon has an angular size of only half a degree.

Now, how large is the Sun in the sky? Your friends will almost all immediately guess that it is bigger. Before they damage their eyes by trying the Moon pinch on the Sun, just remind them that a total eclipse is caused by the Moon completely covering the Sun, so the

Total eclipse in northern India, October 24, 1995. [80mm f/8 fluorite refractor and occulting disk, 635 mm focal length, f/8, 1 second, with ISO 400 film. © 1995 Ernie Piini]

Sun must appear no bigger than a pea in the sky as well. It is the brightness of the Moon and especially the Sun that deceives people into overestimating their angular size.

Now that you have collected on your bets and can lead a life of leisure, think about the remarkable coincidence that allows us to have total eclipses of the Sun. The Sun is 400 times the diameter of the Moon, yet it is about 400 times farther from the Earth, so the two appear almost exactly the same size in the sky. It is this geometry that provides us with the unique total eclipses seen on Earth when our Moon just barely covers the face of the Sun. If the Moon, 2,160 miles (3,476 kilometers) in diameter, were 169 miles (273 kilometers) smaller than it is, or if it were farther away so that it appeared smaller, people on Earth would never see a total eclipse.[1]

It is amazing that there are total eclipses of the Sun at all. As it is, total eclipses can just barely happen. The Sun is not always exactly the same angular size in the sky. The reason is that the Earth's orbit is not circular but elliptical, so the Earth's distance from the Sun varies. When the Earth is closest to the Sun (early January),[2] the Sun's disk is slightly larger in angular diameter, and it is harder for the Moon to cover the Sun to create a total eclipse.

An even more powerful factor is the Moon's elliptical orbit around the Earth. When the Moon is its average distance from the Earth or farther, its disk is too small to occult the Sun completely. In the midst of such an eclipse, a circle of brilliant sunlight surrounds the Moon,

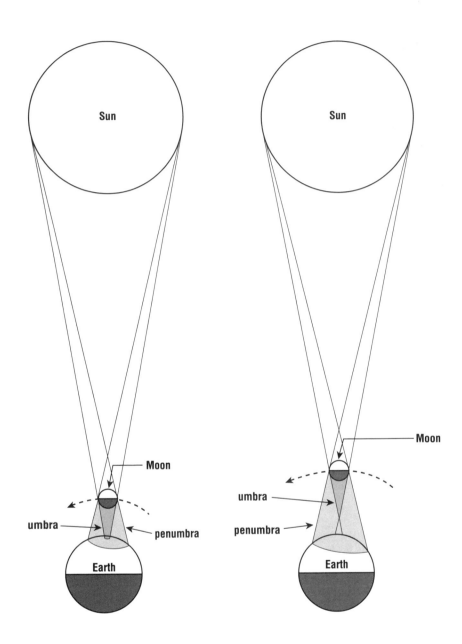

Configuration of a total solar eclipse *(left)* and an annular eclipse *(right)*. Within the Moon's umbra (dark converging cone), the entire surface of the Sun is blocked from view. In the penumbra (lighter diverging cone), a fraction of the sunlight is blocked, resulting in a partial eclipse. When the Moon's umbra ends in space *(right)*, a total eclipse does not occur. Projecting the cone through the tip of the umbra onto the Earth's surface defines the region in which an annular eclipse is seen.

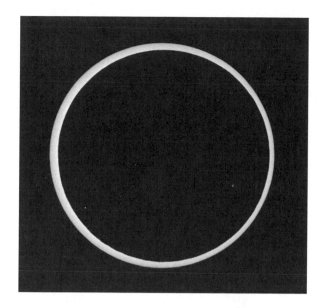

Annular eclipse of May 10, 1994, from Toledo, Ohio. [105 mm f/6 AstroPhysics refractor with 2.5x barlow and glass solar filter, 1500 mm focal length, f/14, 1/125 second, with ISO 100 film. © 1994 Fred Espenak]

giving the event a ringlike appearance; hence the name *annular* eclipse (from the Latin *annulus*, meaning ring).

Because the angular diameter of the Moon is smaller than the angular diameter of the Sun on the average, annular eclipses are more frequent than total eclipses.

But the Moon does not just dangle motionless in front of the Sun. It is in orbit around the Earth. It catches up with and passes the Sun's

Diameters and Distances

	Sun	Moon	Earth
Diameter			
Miles	864,989	2,160	7,927
Kilometers	1,392,000	3,476	12,756
Mean distance from Earth			
Miles	92,960,200	238,870	—
Kilometers	149,598,000	384,400	—

Significance
The Moon's diameter is 1/400 of the Sun's.
The Moon's mean distance is 1/389 of the Sun's.
Thus the Moon and Sun are nearly the *same size* as seen from Earth.

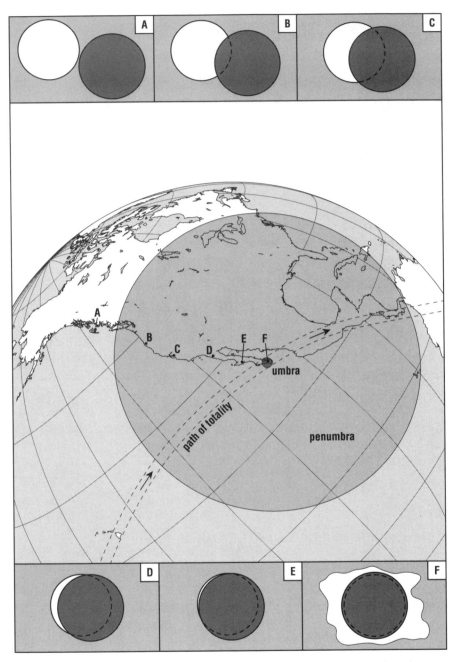

During a total eclipse of the Sun, the tip of the Moon's shadow touches the Earth and the Moon's orbital velocity carries the shadow rapidly eastward. Only along a narrow path is the eclipse total. Regions to the side of the path of totality experience varying degrees of partial eclipse. (This diagram illustrates the total eclipse of July 11, 1991.)

position in the sky about once a month (a period of time derived from this circuit of the Moon). The actual time for the Moon to complete this cycle is 29.53 days, and it is called a synodic month, after the Greek *synodos*, "meeting"—the meeting of the Sun and the Moon. Because the Moon gives off no light of its own and shines

The Shadow of the Moon

	Maximum	Minimum	Mean
Moon's distance from Earth (center to center)			
Miles	252,720	221,470	238,870
Kilometers	406,700	356,400	384,400
Length of Moon's shadow cone (umbra)			
Miles	236,050	228,200	232,120
Kilometers	379,870	367,230	373,540

Significance

On the *average* (in time), the Moon's shadow is too short to reach the Earth. Therefore, total solar eclipses occur *less* often than annular solar eclipses.

Angular Size of the Sun and Moon (as seen from Earth)

	Maximum	Minimum	Mean
Angular diameter of the Sun	32' 31.9"	31' 27.7"	31' 59.3"
Angular diameter of the Moon	33' 31.8"	29' 23.0"	31' 05.3"

Significance

The Moon's angular diameter can *exceed* the Sun's angular diameter by as much as 6.6% (2.1 arc minutes), producing a *total* eclipse of the Sun.

The Sun's angular diameter can *exceed* the Moon's angular diameter by as much as 10.7% (3.1 arc minutes), producing an *annular* eclipse of the Sun.

On the *average* (in time), the Moon's angular diameter is *smaller* than the Sun's angular diameter. Therefore, total solar eclipses occur *less* often than annular solar eclipses.

only by reflected sunlight, its orbit around the Earth changes its angle to the Sun and determines its phase. In 29.53 days, the Moon goes from new moon through full moon and back to new moon again. This period is called a lunar month, a *lunation*. Solar eclipses can take place only at new moon (dark-of-the-moon) and lunar eclipses may occur only at full moon.

So why don't we have an eclipse of the Sun every 29.53 days— every time the Moon passes the Sun's position? The reason is that the Moon's orbit around the Earth is tilted to the Earth's orbit around the Sun by about 5 degrees, so that the Moon usually passes above or below the Sun's position in the sky and cannot block the Sun from our view.

"Danger Zones"

The Moon's tilted orbit crosses the Earth's orbit at two places. Those intersections are called *nodes*. Node is from the Latin word meaning "knot," in the sense of weaving, where two threads are tied together. The point at which the Moon crosses the plane of the Earth's orbit going northward is the ascending node. Going south, the Moon crosses the plane of the Earth's orbit at the descending node.

A solar eclipse can occur only when the Sun is near one of the nodes as the Moon passes.

If the Sun stood motionless in a part of the sky away from the nodes, there would be no eclipses, and you would not be agonizing over this. But the Earth is moving around the Sun, and, as it does so, the Sun appears to move eastward once around the sky, through all the constellations of the zodiac, completing that journey in one year.

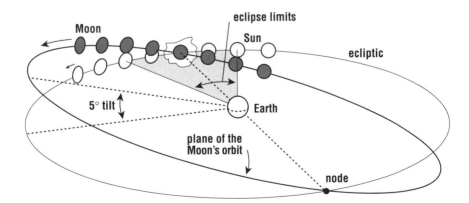

The paths of the Sun and Moon illustrate why eclipses occur only when the Sun is near the intersection (node) where the Moon crosses the ecliptic. The plane of the Moon's orbit is tilted approximately 5° to the ecliptic plane.

Eclipse Limits ("Danger Zones")

	Maximum*	Minimum*	Mean
A solar eclipse of some kind will occur at new moon if the Sun's angular distance from a node of the Moon is less than	18°31'	15°21'	16°56'
A central (total or annular) solar eclipse will occur at new moon if the Sun's angular distance from a node of the Moon is less than	11°50'	9°55'	10°52'

Sun's apparent eastward movement in the star field each day (due to the Earth orbiting the Sun): about 1°

Moon's synodic period (from new moon to new moon: the time the Moon takes to complete its eastward circuit of the star field and catch up with the Sun again): 29.53 days

Significance

The Sun cannot pass a node of the Moon without at least one solar eclipse occurring, and two are possible. If one occurs, it can be either a partial or a central eclipse (total or annular). If two occur, both will be partials, about one month apart.

The Sun can pass a node without a central solar eclipse (total or annular) occurring.

---

*Limits vary due to changes in the apparent angular size and speed of the Sun and Moon caused by the elliptical orbits of the Earth and Moon.

In that yearly circuit, the Sun must cross the two nodes of the Moon. Think of it as a street intersection at which the Sun does not pause and runs the stop sign every time. It is an accident waiting to happen. When the Sun nears a node, there is the "danger" that the Moon will be coming and—crash!

No. The Moon is 400 times closer to the Earth than the Sun, so the worst—the best—that can happen is that the Moon will pass harmlessly but stunningly right in front of the Sun. The Sun's apparent pathway in the sky is called the *ecliptic* because it is only when the Moon is crossing the ecliptic that eclipses can happen. Thus twice a year, roughly, there is a "danger period," called an eclipse season, when the Sun is crossing the region of the nodes and an eclipse is possible.

The Sun comes tootling up to the node traveling about 1 degree a day.[3] The hot-rod Moon, however, is racing around the sky at about 13 degrees a day. Now, if the Sun and Moon were just dots in the heavens, they would have to meet precisely at a node for an eclipse to occur. But the disks of the Sun and Moon each take up about half a degree in the sky. And the Earth, almost 8,000 miles (12,800 kilometers) across, provides an extended viewing platform. Therefore, the Sun needs only to be *near* a node for the Moon to sideswipe it, briefly "denting" the top or bottom of the Sun's face. That will happen whenever the Sun is within 15⅓ degrees of a node.

An "eclipse alert" begins when the Sun enters the danger zone 15⅓ degrees west of one of the Moon's nodes and does not end until the Sun escapes 15⅓ degrees east of that node. The Sun must traverse 30⅔ degrees. Traveling at 1 degree a day, the Sun will be in the danger zone for about 31 days. But the Moon completes its circuit, going through all its phases, and catches up to the Sun every 29.53 days. It is not possible for the Sun to crawl through the danger zone before the Moon arrives. A solar eclipse *must* occur each time the Sun approaches a node and enters one of these danger zones, about every half year.

In fact, if the Moon nips the Sun at the beginning of a danger zone (properly called an *eclipse limit*), the Sun may still have 30 days of travel left within the zone. But the Moon takes only 29.53 days to orbit the Earth and catch up with the Sun again. So it is possible for the Sun to be nipped by the Moon twice during a single node crossing, thereby creating two partial eclipses within a month of one another.

The closer the Sun is to the node when the Moon crosses, the more nearly the Moon will pass over the center of the Sun's face. In fact, if the Sun is within about 10 degrees of the node at the time of the Moon's crossing, a central eclipse will occur somewhere on Earth. Depending on the Moon's distance from the Earth and the Earth's distance from the Sun, this central eclipse will be total or annular.

Nodes on the March

There should be a solar eclipse or two every six months, whenever the Sun crosses one of the Moon's nodes. Actually, the Sun crosses the ascending node of the Moon, then the descending node, and returns to the ascending node in only 346.62 days—the *eclipse year*. Within this eclipse year there are two *eclipse seasons*, intervals of 30 to 37 days as the Sun approaches, crosses, and departs from a node. All eclipses will fall within eclipse seasons. There can be no eclipses out-

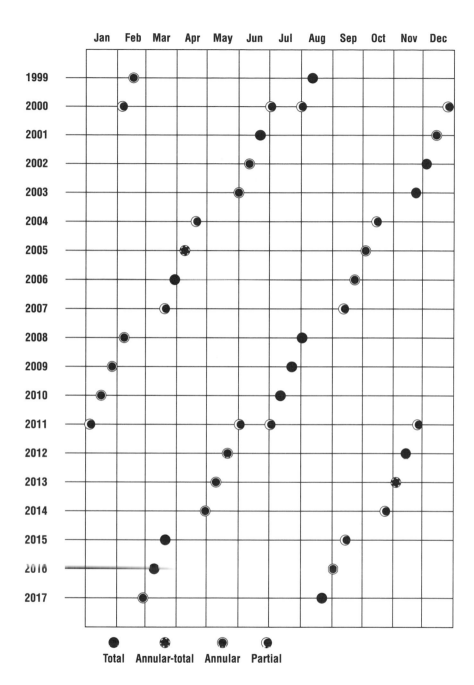

Solar eclipses 1999–2017 plotted to show eclipse seasons. Each calendar year, eclipses occur about 20 days earlier. Consequently, eclipse seasons shift to earlier months in the year.

side this period of time. Because the Sun crosses a node about every 173 days, the eclipse seasons are centered about 173 days apart.

The eclipse year does not correspond to the calendar year of 365.24 days because the nodes have a motion all their own. They are constantly shifting westward along the ecliptic. This regression of the nodes is caused by tidal effects on the Moon's orbit created by the Earth and the Sun. If the nodes did not shift, eclipses would always occur in the same calendar month year after year. If the Sun crossed the nodes in February and August one year to cause eclipses, the eclipses would continue to fall in February and August in succeeding years.

But the eclipse year is 346.62 days, 18.62 days shorter than a calendar year, so each ascending or descending node crossing by the Sun occurs 18.62 days earlier in the calendar year than the previous one of its kind. This migration of the eclipse seasons determines the number of eclipses that may occur each year. One solar eclipse must occur each eclipse season, so there have to be at least two solar eclipses each year (although both may be partial). But because of the width of the eclipse limits—up to 19 days on either side of the Sun's node crossing—and the slowness of the Sun's apparent motion, there can be two solar eclipses at each node passage (both partials). Thus, occasionally there will be four solar eclipses in one calendar year.

There can actually be five. Because the eclipse year lasts 346.62 days, almost 19 days less than a calendar year, if a solar eclipse occurs before or on January 18 (or January 19 in a leap year), that eclipse year could conceivably bring two solar eclipses in January and two more around July. That eclipse year would then end in mid-December, and a new eclipse year would begin in time to provide one final solar eclipse before the end of December. At most, therefore, there can be five solar eclipses in a calendar year.

Heavenly Rhythm

Eclipses, then, are like fresh fruit—available only in season. Ancient peoples who kept written records, such as the Chaldeans from about 747 B.C. on, noticed after decades of observation that eclipses happen only at certain times of the year. These eclipse seasons are separated from one another by either five or six new moons.

From the expanse of their New Babylonian Empire in the Middle East, the Chaldeans could see only about half the lunar eclipses and only a small fraction of the solar eclipses, so, for them, the eclipse seasons were not periods during which one to three eclipses would necessarily occur. Instead, the eclipse seasons were times of "danger"

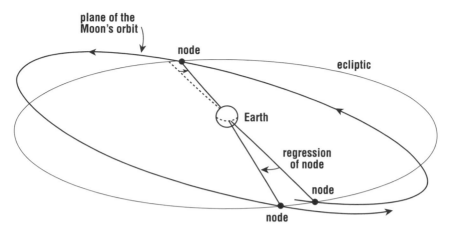

Each time the Moon completes an orbit around the Earth, it crosses the Earth's orbit at a point west of the previous node. Each year the nodes regress 19.4°, making a complete revolution in 18.61 years.

when an eclipse was possible. One of the two great celestial lights might be partially or totally darkened. The Chaldeans did not realize that the invisibility of an eclipse simply meant that it was occurring somewhere else on Earth.

As time passed and they accumulated more records, the Chaldeans and other ancient peoples recognized that a specific eclipse occurred a precise number of days after a previous eclipse and before a subsequent one. Eclipses had a long-term rhythm of their own.

The most famous and, perhaps, most useful of these eclipse rhythms was the *saros*, discovered by the Chaldeans and inscribed on clay tablets in their cuneiform writing.[4] The Chaldeans noticed that 6,585 days (18 years 11 days) after virtually every lunar eclipse, there was another very similar one. If the first was total, the next was almost always total. And these eclipses, separated by 18 years, occurred in the same part of the sky, as if they were related to one another.

In a sense, they were. Imagine that a total solar eclipse occurs one day at the Moon's descending node. After 6,585 days, the Moon has completed 223 lunations (synodic periods) of 29.53 days each and returned to new moon at that same node. In that same period of 6,585 days, the Sun has endured 19 eclipse years of 346.62 days each and has returned to the descending node, forcing another solar eclipse to occur. And because 6,585 days is very close to an even 18 calendar years, this solar eclipse occurs at the same season of the year as its predecessor, and with the Sun very close to the same position in

Solar Eclipses Outnumber Lunar Eclipses

There are more solar than lunar eclipses. This realization usually comes as a surprise because most people have seen a lunar eclipse, while relatively few have seen an eclipse of the Sun. The reason for this disparity is simple. When the Moon passes into the shadow of the Earth to create a lunar eclipse, the event is seen wherever the Moon is in view, which includes half the planet. Actually, a lunar eclipse is seen from more than half the planet because during the course of a lunar eclipse (up to 4 hours), the Earth rotates so that the Moon comes into view for additional areas.

In contrast, whenever the Moon passes in front of the Sun, the shadow it creates—a solar eclipse—is small and touches only a tiny portion of the surface of the Earth. On the average, your house will be visited by a total eclipse of the Sun only once in about 375 years.*

To be touched by the dark shadow (umbra) of the Moon is quite rare. But from either side of the path of a total eclipse, stretching northward and southward 2,000 miles (3,200 kilometers) and sometimes more, an observer sees the Sun partially eclipsed. Even so, this band of partial eclipse covers a much smaller fraction of the Earth's surface than a lunar eclipse. So more people have seen lunar eclipses than partial solar eclipses, and only a tiny fraction of people, about one in 10,000, have witnessed a total solar eclipse.

Yet, to the unaided eye, solar eclipses are substantially more frequent than lunar eclipses. Theodor von Oppolzer, in his monumental *Canon of Eclipses*, published shortly after his death in 1887, attempted to compute, with the help of a number of assistants, all eclipses of the Sun and Moon from 1208 B.C. to A.D. 2161. He cataloged 8,000 solar eclipses and 5,200 lunar eclipses. He found about three solar eclipses for every two lunar eclipses.

This ratio can be misleading, however. Oppolzer counted all solar eclipses, whether they were total (the Moon's *umbra* touches the Earth) or partial (only the Moon's *penumbra* touches the Earth). But for lunar eclipses, Oppolzer counted only those in which the Moon was totally or partially immersed in the Earth's *umbra*. He did not count *penumbral* lunar eclipses because they are virtually unnoticeable. If he had included penumbral lunar eclipses in his census so as to compare the number of all forms of solar and

---

the zodiac that it occupied at the eclipse 18 years earlier. Even though 18 years have intervened, these two eclipses certainly seem to be relatives.

All the more so because another lunar cycle crucial to eclipses has a multiple that also adds up to 6,585 days. That cycle is the *anomalistic month*—the time it takes the Moon in its elliptical orbit around the Earth to go from *perigee* (closest to Earth) to *apogee* (farthest) and back

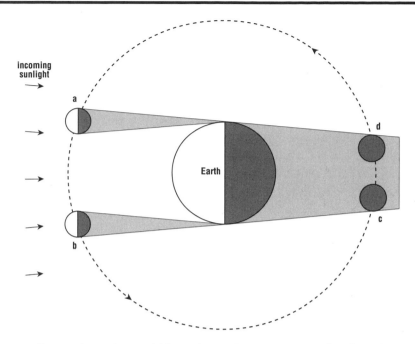

incoming
sunlight
→

a

→

d

Earth

→

→

b

c

→

lunar eclipses, the ratio would have been close to even, with solar eclipses barely prevailing.

The reason why solar eclipses slightly outnumber lunar eclipses is most easily visualized if you imagine looking down on the Sun-Earth-Moon system and if you start by considering only total solar and total lunar eclipses.

The Moon will be totally eclipsed whenever it passes into the shadow of the Earth—between c and d on the diagram. At the Moon's average distance from Earth, that shadow is about 2.7 times the Moon's diameter. But there will be an eclipse of the Sun whenever the Moon passes between the Earth and Sun—between points a and b. The distance between a and b is longer than between c and d, so total solar eclipses must occur slightly more often.

*Jean Meeus: "The Frequency of Total and Annular Solar Eclipses at a Given Place," *Journal of the British Astronomical Association,* volume 92, April 1982, pages 124–126.

to *perigee.*[5] When the Moon is near perigee, its angular size in the sky is just slightly larger, creating eclipses that are total rather than annular. The anomalistic month is 27.55 days long and 239 of these cycles add up to 6,585 days. If the previous eclipse occurred with the Moon at perigee, the new eclipse will occur with the Moon near perigee, providing another total eclipse.

So, after a time interval of 6,585 days, the geometrical configura-

Frequency of Solar Eclipses

Solar eclipses by types (average over 4,530 years):
Total   26.9%
Annular   33.2%
Annular/Total   4.8%
Partial   35.2%

Solar eclipses outnumber lunar eclipses almost 3 to 2 (excluding penumbral eclipses, which are seldom detectable visually)

Annular eclipses outnumber total eclipses about 5 to 4

Solar eclipses per century (average over 4,530 years): 237.8

Number of solar eclipses per year:
Minimum: 2 (either 2 partial or 1 partial and 1 central [total or annular])
Maximum: 5 (4 partial and 1 central)

Number of *total* solar eclipses per year: 0, 1, or 2

Number of solar and lunar eclipses per year:
Minimum: 2 (2 solar and 0 lunar [excluding penumbral eclipses])
Maximum: 7 (4 solar and 3 lunar or 5 solar and 2 lunar)

Recent years with only 2 solar eclipses: 1998, 1999, 2001, 2002

Years in which 5 solar eclipses occur: 1805, 1935, 2206, 2709

Maximum diameter of the Moon's shadow cone (umbra) on Earth during a total eclipse: 170 miles (273 kilometers)

Maximum diameter of the Moon's antumbra on Earth during an annular eclipse: 232 miles (374 kilometers)

---

tion of the Sun, Moon, and Earth is almost exactly repeated. Another solar eclipse occurs at the same node, at the same season, in the same part of the sky. If it was a total eclipse before, it will almost certainly be a total eclipse again.

Take the date of any solar or lunar eclipse and add 6,585.32 days to it and you will accurately predict a subsequent eclipse of the same kind that will closely resemble the one 18 years earlier. Take the date of *every* solar and lunar eclipse that occurs and keep on adding 6,585.32 days to it and you will have, with rare exceptions, a quite reliable list of future eclipses.

Flaws in the Rhythm

But these eclipses 6,585 days apart are not identical twins. The match between unrelated periods—the Moon's cycle of phases, the Sun's eclipse year cycle, and the Moon's cycle from perigee to perigee—is not perfect.

223 synodic periods of the Moon (lunations) at 29.5306 days each	=	6,585.32 days
19 eclipse years of the Sun at 346.6201 days each	=	6,585.78 days
239 anomalistic months of the Moon (revolution from perigee to perigee) at 27.55455 days each	=	6,585.54 days

These synchronizing cycles are out of step with one another by fractions of a day. And those fractions of a day have their consequences.

Consider first that 223 synodic periods of the Moon amount to 6,585.32 days or, in calendar years, 18 years 11⅓ days (18 years 10⅓ days if 5 leap years intervene). Because the saros period is 18 years 11⅓ days, each subsequent eclipse occurs about one-third of the way around the world westward from the one before it. For a lunar eclipse, visible to half the planet, this westward shift usually does not push the eclipse out of view. For a solar eclipse, however, visible over a narrow swath of the Earth's surface, the subsequent eclipse in that saros series would almost never be visible from the initial site. The eclipse would be happening in an entirely different part of the world.

After three saros cycles—54 years 34 days—the eclipse would be back to its original longitude, but it would have shifted, on the average, about 600 miles (1,000 kilometers) northward or southward, taking totality and even major partiality out of view for the original observer.[6]

A saros period truly does predict with accuracy that a solar eclipse will happen, but it would have been extremely hard for an ancient astronomer to confirm that the predicted eclipse had taken place and thus difficult for the astronomer or those he served to retain confidence in his solar eclipse predictions.

The Evolving Saros

Now consider that 223 lunations (synodic periods) amount to 6,585.32 days, while 19 eclipse years of the Sun consume 6,585.78

Total Eclipses—Duration of Totality

Longest duration (theoretical):   7 minutes 31 seconds

Longest duration in 10,000 years*: 7m 29s — July 16, 2186 A.D.

Longest duration by millennium:

3000 B.C. to 2001 B.C.:	7m 21s — May 16, 2231 B.C.
2000 B.C. to 1001 B.C.:	7m 05s — July 3, 1443 B.C.
1000 B.C. to 1 B.C.:	7m 28s — June 15, 744 B.C.
1 A.D. to 1000 A.D.:	7m 24s — June 27, 363 A.D.
1001 A.D. to 2000 A.D.:	7m 20s — June 9, 1062 A.D.
2001 A.D. to 3000 A.D.:	7m 29s — July 16, 2186 A.D.
3001 A.D. to 4000 A.D.:	7m 18s — July 24, 3991 A.D.
4001 A.D. to 5000 A.D.:	7m 12s — August 4, 4009 A.D.

Longest duration of the 20th century (6 minutes or longer)[†]:
   6m 29s — May 18, 1901
   6m 20s — September 9, 1904
   6m 51s — May 29, 1919
   7m 04s — June 8, 1937
   7m 08s — June 20, 1955
   7m 04s — June 30, 1973
   6m 53s — July 11, 1991

Longest duration of the 21st century (6 minutes or longer)[†]:
   6m 39s — July 22, 2009
   6m 23s — August 2, 2027
   6m 06s — August 12, 2045
   6m 06s — May 22, 2096

Number of eclipses with 7 minutes or more of totality in 21st century: 0

Number of eclipses with 7 minutes or more of totality from July 1,
   1098 A.D. to June 8, 1937 A.D. (839 years): 0

Note: All solar eclipses with long durations of totality have dates centered around July 4, the mean date for the Earth at aphelion (farthest from the Sun), when the Sun appears slightly smaller in size and is more easily covered by the Moon.

*Based on the 10,000-year period from 3000 B.C. to 7000 A.D.

[†]All the long total eclipses of the 20th and 21st centuries are members of saros series 136 except for September 9, 1904 and May 22, 2096.

---

days. The difference after 18 years 11⅓ days is 0.46 day. The result is that the Sun is not exactly where it was before in relation to the Moon's node. It is 0.477 degree farther west. If the Moon just barely grazed the western part of the Sun before, it will clip a little more the next time. At each return of the saros, the eclipses become larger and

Annular Eclipses—Duration of Annularity

Longest duration (theoretical):	12 minutes 30 seconds
Longest duration in 10,000 years*:	12m 24s — December 6, 150 A.D.

Longest duration by millennium:

3000 B.C. to 2001 B.C.:	11m 02s — November 24, 2037 B.C.
2000 B.C. to 1001 B.C.:	12m 07s — December 12, 1656 B.C.
1000 B.C. to 1 B.C.:	12m 08s — December 22, 178 B.C.
1 A.D. to 1000 A.D.:	12m 24s — December 6, 150 A.D.
1001 A.D. to 2000 A.D.:	12m 09s — December 14, 1955 A.D.
2001 A.D. to 3000 A.D.:	11m 08s — January 15, 2010 A.D.
3001 A.D. to 4000 A.D.:	12m 09s — January 14, 3080 A.D.
4001 A.D. to 5000 A.D.:	11m 08s — January 20, 4885 A.D.

Longest duration of the 20th century (9 minutes or longer)[†]:
11m 01s — November 11, 1901
11m 37s — November 22, 1919
12m 00s — December 2, 1937
12m 09s — December 14, 1955
12m 03s — December 24, 1973
11m 41s — January 4, 1992

Longest duration of the 21st century (9 minutes or longer)[†]:
11m 08s — January 15, 2010
10m 27s — January 26, 2028
9m 42s — February 5, 2046

Note: All solar eclipses with long durations of annularity have dates centered around January 3, the mean date for the Earth at perigee (closest to the Sun), when the Sun appears slightly larger in size and is harder for the Moon to cover.

Note: A long annularity means that the Moon is smaller than usual in apparent size, creating an annular eclipse that is farthest from being a total eclipse. Thus a long annularity will provide less landscape and sky darkening and less likelihood of a glimpse of Baily's Beads.

*Based on the 10,000-year period from 3000 B.C. to 7000 A.D.

[†]All the long annular eclipses of the 20th and 21st centuries are members of saros series 141.

larger partials until the Moon is passing across the center of the Sun's disk, yielding total or annular eclipses. Then, with the passing generations, as the Sun is farther west within the eclipse limit, the eclipses return to partials. Finally, after about 1,300 years, the Sun is no longer within the eclipse limits when the Moon, after 223 lunations, arrives. After about 1,300 years of adding 6,585.32 days to the date of

Erratic Intervals Between Total Solar Eclipses

Location	Dates	Interval
An average town		average of 375 years
Unusually Long Intervals		
London	October 29, 878 May 3, 1715	837 years
Jerusalem	August 2, 1133 August 6, 2241	1,109 years
United States mainland	February 26, 1979 August 21, 2017	38 years
Unusually Short Intervals—Past		
Brisbane, Australia (region of)	April 5, 1856 March 25, 1857	11 ½ months
New Guinea (southern)	June 11, 1983 November 22, 1984	1 ½ years
Kiev, Ukraine (region of)	March 14, 190 B.C. July 17, 188 B.C. October 19, 183 B.C.	3 in 7 ½ years
Sumatra	May 18, 1901 January 14, 1926 May 9, 1929	3 in 28 years
Jerusalem (region of)	March 1, 357 B.C. July 4, 336 B.C. April 2, 303 B.C.	3 in 54 years
Scotland	March 7, 1598 April 8, 1652 August 12, 1654	3 in 56 years
Spain	July 8, 1842 December 22, 1870 May 28, 1900 August 30, 1905	4 in 63 years
Stonehenge, England	May 3, 1715 May 22, 1724	9 years
Yellowstone National Park, U.S.A.	July 29, 1878 January 1, 1889	10 ½ years

Location	Dates	Interval
Oaxaca, Mexico	March 7, 1970	
	July 11, 1991	21 years
Switzerland	May 12, 1706	
	May 22, 1724	18 years
Wichita, Kansas, U.S.A.	September 17, 1811	
	November 30, 1834	23 years

Unusually Short Intervals—Present and Future

Lobito, Angola (just north of)	June 21, 2001	
	December 4, 2002	1½ years
Confluence of Ohio and Mississippi Rivers, U.S.A. (southeast Missouri, southern Illinois, and western Kentucky)	August 21, 2017	
	April 8, 2024	6⅔ years
Turkey (northern)	August 11, 1999	
	March 29, 2006	6⅔ years
Florida panhandle, U.S.A. (Pensacola to Tallahassee)	August 12, 2045	
	March 30, 2052	6⅔ years
Paris	September 3, 2081	
	September 23, 2090	9 years
Antwerp, Belgium	May 25, 2142	
	June 14, 2151	9 years
Fargo, North Dakota, U.S.A.	September 14, 2099	
	October 17, 2153	54 years
New York City	May 1, 2079	
	October 28, 2144	75 years

an eclipse to predict the next, the prediction fails. No eclipse occurs. That saros has died.[7]

Finally, consider that 223 lunations amount to 6,585.32 days while 239 anomalistic months of the Moon (perigee to perigee) total 6,585.54 days. If the Moon was at perigee and hence largest in angular size during one central eclipse, the eclipse must be total. But with each succeeding eclipse in that saros series, the Moon is a little farther from perigee, until finally the Moon's angular size is too small to completely cover the Sun, and eclipses become annular.

For people long ago, the discovery of the saros and other eclipse

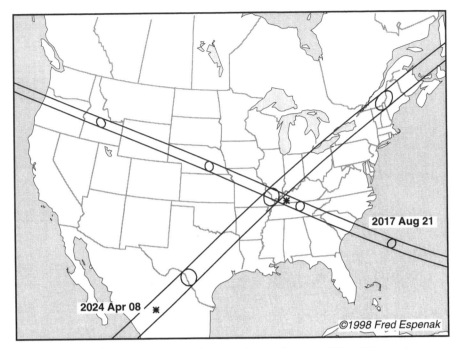

2017 Aug 21

2024 Apr 08

©1998 Fred Espenak

In a period of 6⅔ years, the people near the confluence of the Ohio and Mississippi Rivers in the United States will see two total solar eclipses. [Map and eclipse calculations by Fred Espenak]

cycles probably brought some comfort that eclipses, however dire their interpretation, were part of nature's rhythms. Using these rhythms, it was possible to predict future eclipses without knowing anything about the mechanism that produced them. Even after people understood the causes of eclipses, it was far easier to predict their occurrence by the saros (or other cycles) than to calculate all the factors surrounding the varying motions and apparent sizes of the Sun, Moon, and Earth.

Today, with modern electronic computers to crunch orbits and tilts and wobbles into extremely accurate eclipse predictions, the saros would seem to be an anachronism. Yet it is interesting to view any single total eclipse as a member of an evolving family that had its origin in the distant past, has gradually risen from insignificance to great prominence, and will inevitably decline into oblivion. Chapter 13 explores the genealogy of one of these eclipses.

Creating the Ultimate Eclipse

What conditions would provide the longest total eclipse of the Sun?

First, the Moon should be near maximum angular size, which means it should be near perigee—the point in its orbit when it is closest to Earth. That happens once every 27.55 days (an anomalistic month).

Second, the Sun should be near minimum angular size, which means the Earth should be near aphelion—the point in its orbit when it is farthest from the Sun. That happens once every 365.26 days (an anomalistic year). At present, aphelion occurs in early July.

Third, to prolong the eclipse as much as possible, the Moon's eclipse shadow must be forced to travel as slowly as possible. Here the observer enters the formula. At the time of a solar eclipse, the Moon's shadow is moving about 2,100 miles an hour (3,380 kilometers an hour) with respect to the center of the Earth. But the Earth is rotating from west to east, the same direction that the eclipse shadow travels. At a latitude of 40° north or south of the equator, the surface of the Earth turns at about 790 miles an hour (1,270 kilometers an hour), slowing the shadow's eastward rush by that amount. At the equator, the Earth's surface rotates at 1,040 miles an hour (1,670 kilometers an hour), slowing the shadow's speed to only 1,060 miles an hour (1,710 kilometers an hour), thereby prolonging the duration of totality, Baily's Beads, and all phases of the eclipse.

Finally, to prolong the eclipse just a few seconds more, the Moon should be directly overhead, so that we are using the full radius of the Earth to place ourselves about 4,000 miles (6,400 kilometers) closer to the Moon than the limb of the Earth, thus maximizing the Moon's angular size and therefore its eclipsing power. The latitude where the Moon stands overhead varies with the seasons as the Sun appears to oscillate 23 ½° north and south of the equator. Thus the peak duration of totality rarely occurs at the equator but *always* occurs in the tropics.

The maximum duration of totality for a solar eclipse was calculated by Isabel M. Lewis to be 7 minutes 31 seconds.

Basic Sun and Moon Data Important to Eclipses

	Sun	Moon	Earth
Diameter			
Miles	864,989	2,160	7,927
Kilometers	1,392,000	3,476	12,756
Mean distance from Earth			
Miles	92,960,200	238,870	—
Kilometers	149,598,000	384,400	—

Ratio of Sun's diameter to Moon's diameter: 400.5

Ratio of Sun's mean distance to Moon's mean distance: 389.1

Orbital speed of Moon (mean): 2,290 miles per hour (3,680 kilometers per hour)

Speed through space of Moon's shadow during a solar eclipse (not the same as the Moon's orbital speed because of the orbital motion of the Earth): 2,100 miles per hour (3,380 kilometers per hour)

Orbital eccentricity of Moon: 0.05490

Inclination of Moon's orbit to plane of Earth's orbit (ecliptic) (varies due to tidal effects of Earth and Sun): 5°08' (mean); 5°18' (maximum); 4°59' (minimum)

Regression (westward drift) of Moon's nodes: 19.4° per year

Period for Moon's nodes to regress all the way around its orbit: 18.61 years

3

A Quest to Understand

I look up. Incredible! It is the eye of God. A perfectly black disk, ringed with bright spiky streamers that stretch out in all directions.
—Jack B. Zirker (1984)

Ancient peoples around the world have left monuments and symbols of their reverence for the sky and the results of their efforts to record celestial motions. More than 2,500 years ago, some people could predict or at least warn of the possibility of eclipses, especially lunar eclipses, the fading of the Moon as it passed into the Earth's shadow. And the ability to predict eclipses may go back further still.

Stonehenge

The earliest and most famous of monuments that testify to a people's high level of astronomical knowledge is Stonehenge, near Salisbury in southern England. Awesome, haunting, strangely beautiful, Stonehenge is a permanent record of celestial knowledge in stone. Work was in progress on Stonehenge a generation before construction began on the first pyramids in Egypt. Stonehenge was already abandoned, a desolate mystery, when Moses led the Exodus from Egypt.

The familiar silhouette of archways, now mostly fallen, is the last of four major stages of Stonehenge development that ceased about 1500 B.C. The stones formed a circle of 30 linked archways, approximating the days in a lunar month.[1] Inside this Sarsen Circle was a horseshoe of five even larger freestanding archways, the trilithons, with uprights that weigh up to 50 tons. These massive, shaped boul-

Aerial view of Stonehenge. [English Heritage]

ders were dragged from a quarry 20 miles (32 kilometers) away to codify in stone the discoveries of an earlier people.

The most famous feature of Stonehenge is the line of sight from the center of the monument toward the northeast through an archway in the Sarsen Circle and over the Heel Stone, a 35-ton boulder set upright 245 feet (75 meters) away. From this position, an observer sees the approximate point on the horizon where the Sun rises on the first day of summer, when it is farthest north of the equator and daytime lasts longest. With this alignment and others, the users of Stonehenge could time the beginning of summer and winter with high precision to create an accurate solar calendar, which could have been of great benefit to farmers.

Yet these massive stone archways, raised in the last phase of construction at Stonehenge, show little new in their orientations beyond what archaeologists have found in the original and essential Stonehenge, begun 1,300 years earlier.

The Essential Stonehenge

Stonehenge was begun about 2800 B.C. by a people who had no written language, no wheeled vehicles, no draft animals, and no metal tools. To dig holes in the ground, they used the antlers of deer.

The initial Stonehenge consisted of a circular embankment 350 feet (107 meters) in diameter, four marker stones set in a rectangle, some postholes, and the Heel Stone.[2] The Heel Stone was apparently the first of the great boulders brought to this site as construction commenced. But it may not have stood alone. A similar huge stone stood just to its left as seen from the center of Stonehenge.[3] In that ancient time, the Sun at the beginning of summer probably rose between the famed Heel Stone and its now-vanished companion, and the alignment with sunrise at the summer solstice was probably exact.

For someone standing at the center of Stonehenge, the embankment served to level the horizon of rolling hills. Within the embankment, four stones—the Station Stones—outlined a rectangle within the circle. The sides of this rectangle offered interesting lines of sight. The short side of the rectangle pointed toward the same spot on the horizon that the two Heel Stones framed, the position where the Sun rose farthest north of east, marking the commencement of summer. Facing in the opposite direction along the short side of the rectangle, an observer would see the place where the Sun set farthest south of west, signaling the beginning of winter.

In contrast, the long sides of the rectangle provided alignments for crucial rising and setting positions of the Moon. Looking southeast along the length of the rectangle, an observer was facing the point on the horizon where the summer full moon would rise farthest south.

Viewed from the center of Stonehenge, the Sun rises just to the left of the remaining Heel Stone at the beginning of summer, the longest day of the year. [English Heritage]

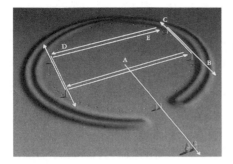

Stonehenge. *Left*: first phase of construction showing alignments *A* and *B*, northernmost sunrise: first day of summer; *C*, southernmost sunset: first day of winter; *D*, southernmost moonrise; *E*, northernmost moonset; *Right*: final phase of construction. [© 1983 Hansen Planetarium]

In the opposite direction, looking northwest, this early astronomer's gaze was led to the spot on the horizon where the winter full moon would set farthest north. These positions marked the north and south limits of the Moon's motion.

The structure of Stonehenge offers additional testimony to its builders' efforts to understand the motion of the Moon. Evidence of small holes near the remaining Heel Stone strongly suggests that the users of Stonehenge observed and marked the excursion of the Moon as much as 5 degrees north and south of the Sun's limit.[4] This motion above and below the Sun's position is caused by the tilt of the Moon's orbit to the Earth's path around the Sun. Because of this tilt, the Moon does not pass directly in front of the Sun (a solar eclipse) or directly into the Earth's shadow (a lunar eclipse) each month.

Because the builders of Stonehenge had discovered and accurately recorded the range in the rising and setting positions of the Sun and Moon and had built a monument that marked these positions with precision, they may have been able to recognize when the Moon was on course to intercept the position of the Sun, to cause a solar eclipse. Perhaps they could tell when the Moon was headed for a position directly opposite the Sun, which would carry it into the shadow of the Earth for a lunar eclipse. They almost certainly could not predict where or what kind of solar eclipse would be seen, but they might have been able to warn that on a particular day or night, an eclipse of the Sun or Moon was *possible*.

In the last phase of building at Stonehenge, two concentric circles of holes were dug just outside the Sarsen Circle—one with 30 holes and the other with 29. These circles reinforce the evidence that astronomers at Stonehenge were counting off the 29½-day cycle of lunar phases, from new moon to full moon and back to new moon again. Eclipses of the Sun can take place only at new moon; lunar eclipses can occur only at full moon. If indeed the lunar phasing cycle was watched carefully, perhaps some ancient genius noticed a periodicity in eclipses as well. With a knowledge of that period, that early astronomer could have converted a mere warning of a possible eclipse into a prediction of a likely eclipse, especially for lunar eclipses, which are visible over half the Earth.[5]

The builders of Stonehenge left no written records of their objectives or results, so we must judge from the monument and its alignments what they knew. Whatever that was, they thought it so worth celebrating that the rulers and apparently the common people were willing to devote vast amounts of time, physical effort, and ingenuity to raising a lasting monument of great size, precision, and beauty.

China

A frequently recounted Chinese story says that Hsi and Ho, the court astronomers, got drunk and neglected their duties so that they failed to predict (or react to) an eclipse of the Sun. For this, the emperor had them executed. So much for negligent astronomers.

If this story were an account of an actual event, the dynasty mentioned would place the eclipse somewhere between 2159 and 1948 B.C., making it by far the oldest solar eclipse recorded in history. But all serious attempts to identify one particular eclipse as the source of this story have been abandoned as scholars have recognized that the episode is mythological.

In ancient Chinese literature, Hsi-Ho is not two persons but a single mythological being who is sometimes the mother of the Sun and at other times the chariot driver for the Sun. Later, in the *Shu Ching* (Historical Classic), parts of which may date from as early as the seventh or sixth century B.C., this single character is split, not into two, but into six. In the *Shu Ching* story, the legendary Chinese emperor Yao commissions the eldest of the Hsi and Ho brothers "to calculate and delineate the sun, moon, the stars, and the zodiacal markers; and so to deliver respectfully the seasons to the people."[6] In further orders, he sends a younger Hsi brother to the east and another to the south; he orders a younger Ho brother to the west and another to the north. Each is responsible for a portion of the rhythms of the days and seasons, to turn the Sun back at the solstices and to keep it moving at the equinoxes.

These mythological magicians are always charged with the prevention of eclipses—hence the story that appears later in the *Shu*

The Hsi and Ho brothers
receive their orders
from Emperor Yao to
organize the calendar.

Ching about the emperor's anger with his servants for failing to *prevent* an eclipse, not just predict or respond ceremonially to it. The story appears in a chapter that is an exhortation by the Prince of Yin, commander of the armies, to government officials to fulfill their duties to the administration, thereby making the emperor "entirely intelligent." If anyone neglects this requirement, "the country has regular punishments for you."

> Now here are Hsi and Ho. They have entirely subverted their virtue, and are sunk and lost in wine. They have violated the duties of their office, and left their posts. They have been the first to allow the regulations of heaven to get into disorder, putting far from them their proper business. On the first day of the last month of autumn, the sun and moon did not meet harmoniously in Fang. The blind musicians beat their drums; the inferior officers and common people bustled and ran

about. Hsi and Ho, however, as if they were mere personators of the dead in their offices, heard nothing and knew nothing;—so stupidly went they astray from their duty in the matter of the heavenly appearances, and rendering themselves liable to the death appointed by the former kings. The statutes of government say, "When they anticipate the time, let them be put to death without mercy; when they are behind the time, let them be put to death without mercy.[7]

We never hear whether Hsi and Ho were ever tracked down and executed.

The story of Hsi and Ho as drunken astronomers was a myth. But the myth did come true in a sense about 33 centuries later. Chinese history records that in A.D. 1202, for the second time in four years, the chief court astronomer made an eclipse forecast that was not as accurate as predictions from people with no official scientific credentials or status. "The astronomical officials were found guilty of negligence and severely punished."[8]

The earliest Chinese word for eclipse, *shih*, means "to eat" and referred to the gradual disappearance of the Sun or Moon as if it were eaten by a celestial dragon.[9] The Chinese were early in recording eclipses but late in recognizing their cause. Not until the third or fourth century A.D. did they understand solar and lunar eclipses well enough to be able to predict them accurately.

The Maya

In the New World, there were ancient people who, like the Chaldeans and the Chinese, used writing to record eclipses and from these records detected a rhythm by which they could predict them or at least warn of their likelihood. Those people were the Maya, and we know of their achievement through one of their books—one of only four that survived the Spanish conquest and its zealous destruction of the religious beliefs of the native peoples.

All that we know of Maya accomplishments in recognizing the patterns of eclipses comes from the Dresden Codex, written in hieroglyphs and pictures in color paints on processed tree bark with pages that open and shut in accordion folds. The book dates from the eleventh century A.D. and is probably a copy of an older work.

We can only wonder what was lost when the conquering Spaniards destroyed by the thousands the books of the Maya and other Mesoamerican peoples. What remains is impressive enough. The Maya realized that discernible eclipses occur at intervals of five or six lunar months. Five or six full moons after a lunar eclipse, there was

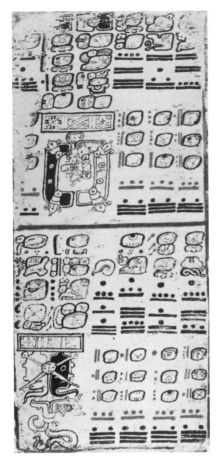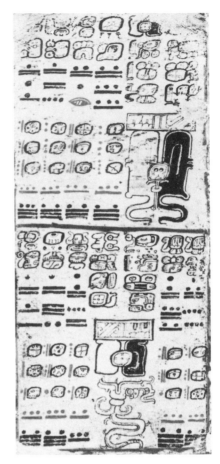

Portion of the Maya solar eclipse prediction tables from the Dresden Codex: at the bottom are the day counts that lead up to a solar eclipse, indicated, *bottom right*, by a serpent swallowing a symbol for the Sun. [American Philosophical Society]

the *possibility* of another lunar eclipse. Five or six new moons after a solar eclipse, another solar eclipse was *possible*.

The Maya had discovered in practical, observable terms the approximate length of the eclipse year, 346.62 days, and the eclipse half year of 173.31 days. The interval for one complete set of lunar phases is 29.53 days. Six lunations amount to approximately 177.18 days, close enough to the eclipse half year (173.31 days) so that there is the "danger" of an eclipse at every sixth new or full moon, but not a certainty. After another six lunar months, the passing days have amounted to 354.36, nearly 8 days *too long* to coincide with the Sun's passage by the Moon's node. An eclipse is less likely. As the error mounts, the need increases to substitute a five-lunar-month cycle into the prediction system rather than the standard six-lunar-month count.

Some great genius must have noticed after recording a sizable number of eclipses that major eclipses were occurring only at intervals of 177 days (6 lunar months) or 148 days (5 lunar months). Using the date of an observed solar or lunar eclipse, it would then have been possible to predict the likelihood of another eclipse, even though in some cases an eclipse would not occur and in others it would not be visible from Mesoamerica.

In the Dresden Codex there are eight pages with a variety of pictures representing an eclipse. Each depiction is different, but most show the glyph for the Sun against a background half white and half black. In two of the pictures, the Sun and background are being swallowed by a serpent. Leading up to each picture is a sequence of numbers: a series of 177s ending with a 148. Each sequence adds up to the number of days in well-known three- to five-year eclipse cycles. At the end of each burst of numbers stands the giant, haunting symbol of an eclipse.

Astronomer-anthropologist Anthony F. Aveni notes that "the reduction of a complex cosmic cycle to a pair of numbers was a feat equivalent to those of Newton or Einstein and for its time must have represented a great triumph over the forces of nature."[10]

From the Maya, we have the numbers that demonstrate one of the greatest of their many discoveries about the rhythms of the sky, but we have no account of the emotion the astronomer-priests or the common folk felt when they observed an eclipse. Perhaps the closest we can come is a passage in the Florentine Codex of the Aztecs, who inherited and used the Mesoamerican calendar but apparently knew little of the astronomy discovered by the Maya a thousand years and more before.

When the people see this, they then raise a tumult. And a great fear taketh them, and then the women weep aloud. And the men cry out, [at the same time] striking their mouths with [the palms of] their hands. And everywhere great shouts and cries and howls were raised. . . . And they said: "If the sun becometh completely eclipsed, nevermore will he give light; eternal darkness will fall, and the demons will come down. They will come to eat us!"[11]

4

Eclipses in Mythology

[T]here was at the same time something in its singular and wonderful appearance that was appalling: and I can readily imagine that uncivilised nations may occasionally have become alarmed and terrified at such an object . . .

—Francis Baily (1842)

Two great lights brighten the heavens. Life depends on them. The disappearance of one threatens the order of the universe and life itself. Through the ages, most cultures responded to eclipses of the Sun and Moon with folklore to explain and endure the eerie events.

Solar eclipse mythology might be divided into several themes, and each of these themes is found scattered throughout the world:

- A celestial being (usually a monster) attempts to destroy the Sun.
- The Sun fights with its lover the Moon.
- The Sun and Moon make love and discreetly hide themselves in darkness.
- The Sun-god grows angry, sad, sick, or neglectful.[1]

Within these myths is a great truth. The harmony and well-being of Earth is dependent on the Sun and Moon. Abstract science cannot convey this profound realization as powerfully as a myth in which the celestial bodies come to life.

The Sun for Lunch

Most often in mythology a solar eclipse is considered to be a battle between the Sun and the spirits of darkness. The fate of Earth and its inhabitants hangs in the balance. With so much at stake, the people enduring an eclipse were anxious to help the Sun in this struggle if they could.

In the mythology of the Norse tribes, Loki, an evil enchanter, is put in chains by the gods. In revenge, he creates giants in the shape of wolves. The mightiest is Mânagarmer (Moon-Wolf, also called Hati), who causes a lunar eclipse by swallowing the Moon. Sköll, another of these wolflike giants, follows the Sun, always seeking a chance to devour it. Old French and German expressions and incantations echo this belief: "God protect the Moon from wolves."

In India, Rāhu is one of the Asurases, demons who are the elder brothers of the gods. The Asurases and the gods fight over the possession of Lakṣmī, goddess of wealth and beauty, and the possession of ambrosia. The Asurases had captured the ambrosia and Rāhu was drinking it when the god Nārāyana caught up with him, threw his discus, and sliced Rāhu in half. That is why Rāhu has a head but no body. The head flew off into the sky where it attacks the Sun and Moon, swallowing them to cause an eclipse. The severed lower body of Rāhu is Ketu and has become the constellations.

Buddhism placed the Buddha in a position to correct the celestial terrorism of Rāhu. In the midst of an eclipse, the Sun or Moon cries out to the Buddha for help. "Rāhu," says the Buddha, "let go of the [Sun], for the Buddhas pity the world." Rāhu departs in terror, fearing that, if he harms the Sun or the Moon, the Buddha will cause his head to shatter into seven parts.[2]

The Indian tale of Rāhu spread eastward into China and northward into Mongolia and eastern Siberia, where the monster's name became Arakho or Alkha. The Buryat people, living east of Lake Baikal, told of Alkha, a monster who continually pursued and swallowed the Sun and Moon. Finally, the gods were exasperated by the repeated darkening of the world and cut Alkha in two. His lower body fell to Earth, but the upper portion lives on and continues to haunt the sky. This is why the Sun and Moon still disappear from time to time: Alkha swallows them. But they soon reappear because Alkha's body cannot retain them. When eclipses are in progress, say the Buryats, the Sun and Moon pray for help, and the people respond by screaming and by throwing stones and shooting weapons into the sky to scare away the monster.

Another Buryat myth says that the eclipse-maker is Arakho, a beast who formerly lived on Earth. In those days long ago, the people were quite hairy. Arakho roamed the Earth eating the hair off their bodies until the people had become the nearly hairless creatures they are today. This annoyed the gods, who chopped Arakho in two. Thus

Arakho no longer grazes on human hair. Instead, the upper portion of his body, which is still alive, eats the Sun and Moon, causing eclipses.

Rāhu appears again in Indonesia and Polynesia as Kala Rau, all head and no body, who eats the Sun, burns his tongue, and spits the Sun out.

Ancient Egypt produced a tale of the black pig, the evil god Set in disguise, who leaped into the eye of Horus, the Sun god. The story becomes confused before we learn how the eye was healed, but it may have been the work of Thout, the Moon, who regulates such disturbances as eclipses and is also the healer of eyes.[3]

In northwestern France (Ille-et-Vilaine), the people recognized that eclipses occur when the Moon blocks the Sun from view. For them, the Moon's aggression took a different form. If the Moon were to cover the Sun entirely, it would stick in that position so that the Sun would never shine again.[4]

Egyptian emblem of the winged Sun at the top of the Gateway of Ptolemy at Karnak. [William Tyler Olcott: *Star Lore of All Ages*]

Drawing of the corona by Samuel P. Langley from the top of Pike's Peak, July 29, 1878. The elongated corona resembles the Egyptian emblem of the winged Sun. [Mabel Loomis Todd: *Total Eclipses of the Sun*]

From around the world come stories of many different monsters intent on devouring the Sun and Moon. In China, it was a heavenly dog who causes eclipses by eating the Sun. In South America, the Mataguaya Indians of the pampas saw eclipses as a great bird with wings outspread, assailing the Sun or Moon. In Armenian mythology, eclipses were the work of dragons who sought to swallow the Sun and Moon. By contrast, another Armenian myth says that a sorcerer can stop the Sun or Moon in their courses, deprive them of light, and even force them down from the skies. Despite the Moon's size, once it has been brought down to Earth, the sorcerer can milk it like a cow.[5]

On rare occasions in mythology, the Sun and Moon are not totally innocent victims. In a variant Hindu myth, the Sun and Moon once borrowed money from a member of the savage Ḍom tribe and failed to pay it back. In retribution, the Ḍom occasionally devours the two heavenly bodies.[6]

The Original Black Holes

In two instances, the hungry monster who swallows the Sun or Moon becomes quite scientifically sophisticated in character. A western Armenian myth, said to be borrowed from the Persians, tells of

The solar corona of February 26, 1998 from Oranjestad, Aruba. [3.5-inch Questar, 1440 mm focal length at f/16, 1 second, with ISO 400 film, © 1994 by Ken Willcox]

two dark bodies, the children of a primeval ox. These dark bodies orbit the Earth closer than the Sun and Moon. Occasionally they pass in front of the Sun or Moon and thereby cause an eclipse.[7]

Still more remarkable is a Hindu myth that speaks of the Navagrahas, "the nine seizers." These nine "planets" that wander through the star field include the usual seven familiar to the Greeks—Sun, Moon, Mercury, Venus, Mars, Jupiter, and Saturn—plus Rāhu and Ketu, "regarded as the ascending and descending nodes" of the Moon, the shifting points in the sky where the Moon crosses the apparent path of the Sun.[8] Thus, quite correctly, the Sun would be at risk of an eclipse whenever it passed by Rāhu or Ketu.

Buddhism carried Rāhu and Ketu from India to China in the first century A.D., where they became Lo-Hou and Chi-Tu. They were imagined as two invisible planets positioned at the nodes in the Moon's path: Lo-Hou at the ascending node and Chi-Tu at the descending node. These "dark stars" were numbered among the planets and were considered to be the cause of eclipses.[9]

Love, Marriage, and Domestic Violence

A Germanic myth explained eclipses differently. The male Moon married the female Sun. But the cold Moon could not satisfy the passion of his fiery bride. He wanted to go to sleep instead. The Sun and Moon made a bet: whoever awoke first would rule the day. The Moon promptly fell asleep, but the Sun, still irritated, awoke at 2 A.M. and lit up the world. The day was hers; the Moon received the night. The Sun swore she would never spend the night with the Moon again, but she was soon sorry. And the Moon was irresistibly drawn to his bride. When the two come together, there is a solar eclipse, but only briefly. The Sun and Moon begin to reproach one another and fall to quarreling. Soon they go their separate ways, the Sun blood-red with anger.

It was not always a fight between the Sun and Moon that caused an eclipse. Sometimes it was love, and modesty. The Tlingit Indians of the Pacific coast in northern Canada explained a solar eclipse as the Moon-wife's visit to her husband. Across the continent, in southeastern Canada, the Algonquin Indians also envisioned the Sun and Moon as loving husband and wife. If the Sun is eclipsed, it is because he has taken his child into his arms.[10]

For the Tahitians, the Sun and Moon were lovers whose union creates an eclipse. In that darkness, they lose their way and create the stars in order to light their return.

An Angry Sun

Sometimes, as in a folktale from eastern Transylvania, it is the perversion of mankind that brings on an eclipse. The Sun shudders, turns away in disgust, and covers herself with darkness. Stinking fogs gather. Ghosts appear. Dogs bark strangely and owls scream. Poisonous dews fall from the skies, a danger to man and beast. Neither humankind nor animals should consume water or eat fresh fruit or vegetables. Such beliefs persisted into the nineteenth century. This poisonous dew that supposedly accompanied eclipses could be the source of an outbreak of the plague or other epidemics. If people had to leave their homes, they wrapped a towel around their mouths and noses to strain out the noxious vapors. Clothes caught drying outdoors during a solar eclipse were considered to be infected.

The Germans were not alone in their belief that a solar eclipse brought a dangerous form of precipitation. Eskimos in southwestern Alaska believed that an unclean essence descended to Earth during an eclipse. If it settled on utensils, it would produce sickness. Therefore, when an eclipse began, every Eskimo woman turned all her pots, buckets, and dishes upside down.[11]

Yet it was not always an angry Sun that brought darkness to the Earth. When U.S. Coast Survey scientist George Davidson observed the eclipse of August 7, 1869, from Kohklux, Alaska, he found that the Indians there attributed the eclipse to an illness of the Sun. He had alerted them to the impending eclipse, but they doubted it. Halfway through the partial phase, the Indians and their chief quit work and hid in their houses: "[T]hey looked upon me as the cause of the Sun's being 'very sick and going to bed.' They were thoroughly alarmed, and overwhelmed with an indefinable dread."[12]

Sometimes in mythology, an eclipse is not a monster devouring the Sun, not a sickness of the Sun, not a fight between the Sun and Moon, not even the result of the always abundant sins of mankind. Sometimes an eclipse is what in sports would be called an unforced error. The Bella Coola Indians of the Pacific coast in Canada had a myth that began with a remarkable observational description of the Sun's apparent annual pathway though the sky. The trail of the Sun, they said, is a bridge whose width is the distance between the summer and winter solstices, the northernmost and southernmost positions of the Sun. In the summer, the Sun walks on the right side of this bridge; in the winter, he walks on the left. The solstices are where the Sun sits down. Accompanying the Sun on his journey are three guardians who dance about him. Sometimes the Sun simply drops his torch, and thus an eclipse occurs.[13]

Warding Off the Evil

Corruption and death are a frequent theme of eclipse myths. Evil spirits descend to Earth or emerge from underground during eclipses.

On June 16, 1406, says an enlightened and bemused French chronicler, "between 6 and 7 A.M., there was a truly wonderful eclipse of the Sun which lasted nearly half an hour. It was a great shame to see the people withdrawing to the churches and believing that the world was bound to end. However the event took place, and afterward the astronomers gathered and announced that the occurrence was very strange and portended great evil."[14] (Half an hour is too long for totality, but the right length for conspicuously reduced light surrounding totality.)

For Hindus, the place to be during an eclipse was in the water, especially in the purifying current of the Ganges.[15] This Hindu practice of immersing oneself in water was known to the French philosopher and popularizer of science Bernard Le Bovier de Fontenelle, as recorded in his *Entretiens sur la pluralité des mondes* (Conversations on the Plurality of Worlds) in 1686.

> All over India, they believe that when the sun and moon are eclipsed, the cause is a certain dragon with very black claws which tries to seize those two bodies, wherefore at such times the rivers are seen covered with human heads, the people immersing themselves up to the neck, which they regard as a most devout position, and implore the sun and moon to defend themselves well against the dragon.[16]

A Participatory Event

From around the world come reports of people trying to assist the Sun and Moon against the peril of eclipse. Screaming, crying, and shouting are supposed to encourage the Sun and Moon to escape the clutches of the evil spirit. Historians recorded that Germans watching a lunar eclipse in the Middle Ages chanted in unison, "Win, Moon."

The Sun and Moon were the supreme gods of the Indians of Colombia. When these gods were threatened by an eclipse, they seized their weapons and made warlike sounds on their musical instruments. They also shouted to the gods, promising to mend their ways and work hard. To prove it, they watered their corn and worked furiously with their tools during the eclipse.[17]

People frequently augmented their voices with the clanging of metal pots, pans, and knives. The Chippewa Indians in the northeastern United States and southeastern Canada went even further. Seeing the Sun's light being extinguished, though not by a monster, they shot flaming arrows into the sky, hoping to rekindle the Sun.[18] The Sencis of eastern Peru did the same, but to scare off a savage beast attacking the Sun.

Ethnographers descended on the Kalina tribe in Suriname to collect their folklore and watch their behavior as the total eclipse of June 30, 1973, neared. In Kalina mythology, the Sun and Moon are brothers. They usually get along well together, but occasionally they have sudden and ferocious quarrels that endanger mankind. At such times, it is important to separate the combatants by making a maximum of noise: banging on tools, hollow objects, and instruments. The fading of the Sun or Moon means that one has been knocked unconscious. When this happens, the tribesmen yell, "Wake up, Papa!" Papa, here, is a term of respect, not an indication that the Kalina consider themselves descended from the Sun or Moon. After the 1973 eclipse, invisible because of clouds but noticeable because of darkening, the old women (the pot makers) rounded up the children and used branches to spread white clay all over them, somewhat too vigorously for the children's enjoyment. Then they smeared the women and finally the men from head to foot with white clay. The white clay was the blood of the injured Moon that had dripped onto the ground. It was necessary to wash oneself with the Moon's blood to restore purity in man and whiteness to the Moon. After an hour, the tribesmen washed themselves off in the river.[19]

In ancient Mexico and Central America, the most important god was represented as a plumed serpent. For the Maya, he was Kukulcán; for the Aztecs, Quetzalcóatl. At an eclipse of the Moon and most especially during an eclipse of the Sun, a special snake was killed and eaten.[20]

In prehistoric times, screams, cries, banging noises, and prayer may not have been deemed adequate to ward off eclipses or their effects. In many places around the world, human sacrifice was performed at the appearance of unexpected and confusing sights, such as an eclipse or a comet. Yet few eclipse myths refer to human sacrifice, suggesting that this practice had largely been abandoned before most eclipse myths were preserved. An exception was in Mexico and Central America, where the Spanish invaders saw the Aztecs and their neighbors carry out human sacrifice in the early sixteenth century. For the Aztecs, almost any natural or political event was commemorated with a sacrifice. On the occasion of an eclipse, the Sun was in need of help from people, just as he had constant help from the dog Xolotl (Sho-LOT-uhl). Xolotl was the god of human monstrosities (including twins), so it was humpbacks and dwarfs that were sacrificed to the Sun to help him prevail.[21]

Solar eclipses boded ill for everyone, it seems, except prospectors. In Bohemia, people believed that a solar eclipse would help them find gold.

5

Strange Behavior of Man and Beast

The Sun . . .
In dim eclipse disastrous twilight sheds
On half the nations, and with fear of change
Perplexes monarchs.
                    —John Milton (1667)

The Human Response

One of the most dramatic responses in history to a total solar eclipse is presented by Herodotus, the first Greek historian, writing around 430 B.C.

> [W]ar broke out between the Lydians and the Medes [major powers in Asia Minor], and continued for five years, with various success. In the course of it the Medes gained many victories over the Lydians, and the Lydians also gained many victories over the Medes. . . . As, however, the balance had not inclined in favour of either nation, another combat took place in the sixth year, in the course of which, just as the battle was growing warm, day was on a sudden changed into night. This event had been foretold by Thales, the Milesian, who forewarned the Ionians of it, fixing for it the very year in which it actually took place. The Medes and Lydians, when they observed the change, ceased fighting, and were alike anxious to have terms of peace agreed upon.

A treaty was quickly made and sealed by the marriage of the daughter of the Lydian king to the son of the Median king.[1] This story indicates the awe that ancient people felt when confronted with a total eclipse of the Sun.

During a battle between the Lydians and the Medes on May 28, 585 B.C., a total eclipse of the Sun occurred. It scared the soldiers so badly that they stopped fighting and signed a treaty. [Mabel Loomis Todd: *Total Eclipses of the Sun*]

Modern astronomers, armed with the dates of the kings described in the account and a knowledge of the dates and paths of ancient eclipses, have generally settled upon May 28, 585 B.C., as the eclipse to which the story refers, if Herodotus, given to fanciful embellishment, can be trusted about an event that occurred a century before he was born.[2]

In his account, Herodotus credits Thales with predicting this eclipse. If so, Thales would have been the first person *known* to have calculated a future solar eclipse. Cuneiform writing on clay tablets from the Chaldean (or New Babylonian) Empire dating about two centuries later shows recognition of an 18-year-11-day rhythm in eclipses—the saros. Perhaps Thales borrowed this eclipse rhythm from the Babylonians as it was being developed. However, such a rhythm predicts not just the year but the month and precise day of the eclipse. Yet Herodotus seems amazed that Thales could be accurate to "the very year in which it actually took place." Was Herodotus so surprised that Thales could predict an eclipse accurate to the day that he simply could not believe that degree of precision and used the more conservative "year" instead? That would be out of character for the flamboyant Herodotus. Yet predicting a solar eclipse accurate to a year is not much of a trick, since there are a minimum of two solar eclipses a year. The problem is to predict a total eclipse for a particular location on Earth. Could Thales have accomplished this? It is doubtful.

The saros period is actually 18 years 11⅓ days, so the Earth has spun through an extra 8 hours, so each subsequent eclipse falls about one-third of the way around the world westward from the one before it. Thus successive eclipses in a saros series are almost never visible from the same site.

The eclipse in the same saros series that preceded 585 B.C. occurred on May 18, 603 B.C., with an early-morning path from the northern portion of the Red Sea to the northern tip of the Persian Gulf, about 600 miles (1,000 kilometers) distant from the end of the path of the May 28, 585 B.C. eclipse. Thales could have heard reports of the 603 B.C. eclipse and used it to calculate the date for the 585 B.C. eclipse. But the saros projection would not have told him where the eclipse would be visible. Thales, then, first of the great Greek philosophers, could have warned of the *possibility* of a solar eclipse, but he could not predict from the saros period that it would be visible in Asia Minor. And there is no evidence that he had the celestial knowledge or the mathematics to calculate it from orbital considerations.

Of course the key to appreciating the story of the solar eclipse that stopped a war is the realization that people long ago were stunned by a total eclipse of the Sun and incredulous that someone could predict such an event. Quite often in ancient history, eclipses are reported to have played a decisive role in the turn of events.

Herodotus tells of another turning point in world history that he says hinged on a solar eclipse. Xerxes and his Persian army were about to march from Sardis to Abydos on their advance toward Greece.

> At the moment of departure, the sun suddenly quitted his seat in the heavens, and disappeared, though there were no clouds in sight, but the sky was clear and serene. Day was thus turned into night; whereupon Xerxes, who saw and remarked the prodigy, was seized with alarm, and sending at once for the Magians, inquired of them the meaning of the portent. They replied—"God is foreshadowing to the

Greeks the destruction of their cities; for the sun foretells for them, and the moon for us." So Xerxes, thus instructed, proceeded on his way with great gladness of heart.[3]

To disaster! He reached and burned Athens, but his navy was destroyed by the Greeks and his forces had to withdraw. Twice more Xerxes invaded Greece, but each time his armies were crushed. After his last defeat, his nobles assassinated him.

Xerxes' first march against Greece actually occurred in 480 B.C., but the only major eclipse visible in the region near that date was the total eclipse of February 17, 478 B.C. Thus, the story tells us less about observational astronomy in that era than about the power exercised by eclipses over the minds of men and the effectiveness of their use to heighten the drama of a story.

One final story illustrates the advance of the Greeks from superstitious dread of eclipses to an understanding of what causes them. On August 3, 430 B.C., Pericles and his fleet of 150 warships were about to sail for a raid upon their enemies.

> But at the very moment when the ships were fully manned and Pericles had gone onboard his own trireme, an eclipse of the sun took place, darkness descended and everyone was seized with panic, since they regarded this as a tremendous portent. When Pericles saw that his helmsman was frightened and quite at a loss what to do, he held up his cloak in front of the man's eyes and asked him whether he found this alarming or thought it a terrible omen. When he replied that he did not, Pericles asked, "What is the difference, then, between this and the eclipse, except that the eclipse has been caused by something bigger than my cloak?" This is the story, at any rate, which is told in the schools of philosophy.[4]

The eclipse was a large partial at Athens and annular about 600 miles (1,000 kilometers) to the northeast. This eclipse had also been recorded by Thucydides, without the didactic story, but exhibiting an increased awareness of the cause of eclipses: "The same summer, at the beginning of a new lunar month, the only time by the way at which it appears possible, the sun was eclipsed after noon. After it had assumed the form of a crescent and some stars had come out, it returned to its natural shape."[5]

By 1654, Paris was a center of enlightenment, but on August 12, "at the mere announcement of a total eclipse, a multitude of the inhabitants of Paris hid themselves in deep cellars."[6]

The Shawnee Prophet Uses an Eclipse

Tenskwatawa, the Shawnee Prophet (1775?–1837?), was an important Indian religious leader in Ohio and Indiana in the early nineteenth century. He saw great danger for his people as they increasingly adopted the customs of the European settlers, especially alcohol. He urged them to return to traditional Indian ways and to unite into a single Indian nation under the leadership of his brother Tecumseh to resist the encroachment of white men with their fraudulent treaties.

General William Henry Harrison, later president of the United States, was at that time the governor of Indiana Territory, where the Shawnee Prophet was successfully recruiting converts to his Indian religious revival. Seeking to undermine the credibility of the Shawnee Prophet as a shaman, Harrison urged Indians to demand proof from the Prophet that he could perform miracles. Thinking in biblical terms, Harrison asked if Tenskwatawa could "cause the Sun to stand still, the Moon to alter its course, the rivers to cease to flow, or the dead to rise from their graves."

The followers of the Shawnee Prophet did not need such displays, but Tenskwatawa was a canny politician. He proclaimed that on July 16, 1806, he would blot the Sun from the sky as a sign of his divine powers. Whether he knew of this total eclipse from a British agent or from an almanac is uncertain, but a great many Indians gathered at the Shawnee Prophet's camp as the appointed day dawned clear.

At the proper moment, the Prophet, in full ceremonial regalia, pointed his finger at the Sun, and the eclipse began. When the Prophet called out to the Good Father of the Universe to remove his hand from the face of the Sun, the light gradually returned to the Earth. Response to the Prophet's performance was overwhelming and his fame spread rapidly and widely. Harrison's condescension had backfired, to his embarrassment.

But the westward migration of European settlers was unstoppable. In the Battle of Tippecanoe in 1811, Harrison destroyed the Shawnee Prophet's religious center, killing many Indians, and breaking the power of Tenskwatawa.*

*See especially Laurence A. Marschall: "A Tale of Two Eclipses," *Sky & Telescope*, volume 57, February 1979, pages 116–118.

No wonder then that the sight of a total eclipse on July 29, 1878, had a powerful effect on Native Americans near Fort Sill, Indian Territory (now Oklahoma). A non-Indian described it this way:

It was the grandest sight I ever beheld, but it frightened the Indians badly. Some of them threw themselves upon their knees and invoked the Divine blessing; others flung themselves flat on the ground, face downward; others cried and yelled in frantic excitement and terror. Finally one old fellow stepped from the door of his lodge, pistol in hand, and, fixing his eyes on the darkened Sun, mumbled a few unintelligible words and raising his arm took direct aim at the luminary, fired off his pistol, and after throwing his arms about his head in a series of extraordinary gesticulations retreated to his own quarters. As it happened, that very instant was the conclusion of totality. The Indians beheld the glorious orb of day once more peep forth, and it was unanimously voted that the timely discharge of that pistol was the only thing that drove away the shadow and saved them from . . . the entire extinction of the Sun.[7]

The Animal Response

Noting their own primal response to the daytime darkening of the Sun, people through the ages have been fascinated by the reaction of animals to a total eclipse. Reports go back more than 750 years. In describing the eclipse of June 3, 1239, Ristoro d'Arezzo wrote: "[W]e saw the whole body of the Sun covered step by step . . . and it became night . . . and all the animals and birds were terrified; and the wild beasts could easily be caught . . . because they were bewildered."[8]

As a university student in Portugal, astronomer Christoph Clavius saw the total eclipse of August 21, 1560: "[S]tars appeared in the sky, and (miraculous to behold) the birds fell down from the sky to the ground in terror of such horrid darkness."[9]

In 1706 at Montpellier in southern France, observers reported that "bats flitted about as at the beginning of night. Fowls and pigeons ran precipitately to their roosts." In 1715, the French astronomer Jacques Eugène d'Allonville, Chevalier de Louville, traveled to London for the eclipse and observed that at totality "horses that were laboring or employed on the high roads lay down. They refused to advance."[10]

By 1842, some people were even conducting behavioral experiments on their pets. "An inhabitant of Perpignan [France] purposely kept his dog without food from the evening of the 7th of July. The next morning, at the instant when the total eclipse was going to take place, he threw a piece of bread to the poor animal, which had begun to devour it, when the sun's last rays disappeared. Instantly the dog let the bread fall; nor did he take it up again for two minutes, that is, until the total obscuration had ceased; and then he ate it with great avidity."[11]

William J. S. Lockyer, son of the pioneering solar spectroscopist, traveled to Tonga for an eclipse in 1911. The weather conditions were miserable and the insects numerous and very hungry. He and his colleagues caught only a brief view of the corona through thin clouds, and the scientific results were meager. The only members of his team with good results were those studying animal behavior. The horses did not seem to notice the darkening, but fowl ran home to roost and pigs lay down. Flowers closed. But most memorable of all were the insects, which had been completely silent until the moment of totality and then sang as if it were night. "The noise," recalled Lockyer, "was most impressive, and will remain in my memory as a marked feature of that occasion."[12]

6

Anatomy of the Sun

"And it will come about in that day," declares the Lord God, "that I shall make the Sun go down at noon and make the Earth dark in broad daylight."

—Amos 8:9

At the Core

At this moment, deep in the core of the Sun, the nucleus of a hydrogen atom—a proton—is colliding and fusing with another hydrogen nucleus, and the collisions and fusions proceed until four hydrogen nuclei have become the nucleus of one helium atom. In this nuclear reaction, a tiny amount of mass has been destroyed: not lost, but converted into energy. It is this reaction that powers the Sun and all stars, creates their light and heat, for more than 90 percent of their active lives.

In the Sun, tens of trillions of these reactions take place every *second*. Every second 600 million tons of hydrogen become 596 million tons of helium, and 4 million tons of mass become energy, in accordance with Einstein's famous equation $E = mc^2$. A little bit of mass yields a vast amount of energy.[1]

Even though the Sun is actually losing mass at the rate of 4 million tons every second, this weight-reduction plan is far from a crash diet. At 4 million tons a second, it would take the Sun 14.8 trillion years to consume itself entirely, if it could. But it can't. The heat to run this nuclear fusion comes from the gravitational force of all the mass of the Sun pressing inward on the core, and it is only within this central 25 percent of the Sun's diameter that the temperatures—

up to 27 million degrees Fahrenheit (15 million degrees Celsius)—are hot enough to generate and sustain this reaction.[2]

Over billions of years, the hydrogen at the core is converted into helium until the core is too clogged with helium for hydrogen fusion to continue. It is then that stars begin to die. Our Sun has been shining for 4.6 billion years and has enough hydrogen at its core to continue shining much as it does now for about another 5 billion.

The fiercely hot reaction at the heart of our Sun is concealed from our view by 432,500 miles (696,000 kilometers) of opaque gases. And a good thing too. The principal radiation generated at the Sun's core is not visible light but gamma rays. There would be no life on Earth if the Sun radiated large quantities of this high-energy radiation in our direction.

Fortunately, it does not. High-energy photons radiate outward from the Sun's core but are absorbed in the crush of other atomic particles, which in turn reradiate this energy. But the energy emitted is not the same. With each absorption and emission, some of the photon's original energy is lost. A photon that started as a highly energetic gamma ray, if tracked from absorption to absorption, would gradually become an X ray, then an ultraviolet ray, and then

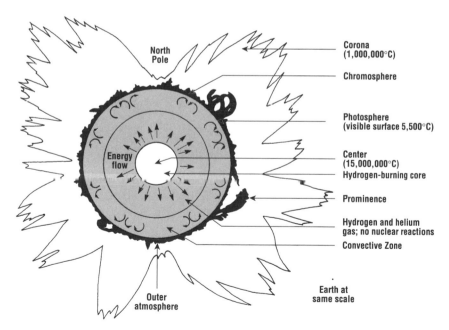

Cross section of the Sun and its atmosphere. [Drawing by Josie Herr after William K. Hartmann: *Cosmic Voyage Through Time and Space*, © 1990 Wadsworth Publishing, Inc., by permission of the publisher]

visible light as it bounced around randomly inside the Sun. At about 130,000 miles (200,000 kilometers) below the Sun's surface, the temperature and density have fallen enough so that energy is conveyed upward less by radiation than by convection—the rise of gases heated by this energy from below.

If energy created at the Sun's core could leave the Sun directly traveling at the speed of light, it would emerge from the Sun's surface in 2 1/3 seconds. But because of the countless absorptions and reemissions in random directions, a typical photon requires 10 million years to reach the Sun's surface. There, at last, it is free to move directly away from the Sun at the speed of light. At that pace, it travels the distance from the Sun to the Earth in 8 1/3 minutes.

Layers Above

All the Sun's internal layers of radiation and convection are hidden from our eyes. What we see of the Sun, when we or the atmosphere provide adequate filters, is an apparent disk, glaring white-hot at a temperature of about 10,000°F (5,500°C). This disk is an optical illusion, since the Sun is gaseous throughout. We are really seeing the layer of the Sun in which the density and ionization of atoms is so great that the gas becomes opaque. This region that provides the Sun with the appearance of a surface is called the *photosphere* ("light sphere"). It is only about 200 miles (300 kilometers) deep. It is there that we notice sunspots, areas of magnetic disturbance on the Sun the size of Earth or Jupiter or even larger, appearing and disappearing and riding along in the photosphere with the Sun's rotation. The sunspots increase and decrease in number over a period of about 11 years.

If we examine the photosphere more closely, we see that it has a mottled appearance, created by rising columns of hot, bright gas surrounded by darker haloes where the gas has cooled and is descending, to be heated again. The photosphere is boiling. These *granulations* of upwelling and downfalling gases are typically 500 miles (800 kilometers) in diameter and last for five or ten minutes. The gases in the granulations, the topmost layer of the convection zone, carry with them magnetic fields from deep within the Sun. As these gases tumble up and down, the magnetic lines of force twist and snap, and this magnetic turbulence governs the behavior of gases in the photosphere and in the solar atmosphere above it.

Above the photosphere is the *chromosphere* ("color sphere"), aptly named for its vibrant reddish color. Seen with a telescope at the rim of the Sun, it looks like a fire-ocean.[3] Its lower levels are cooler than

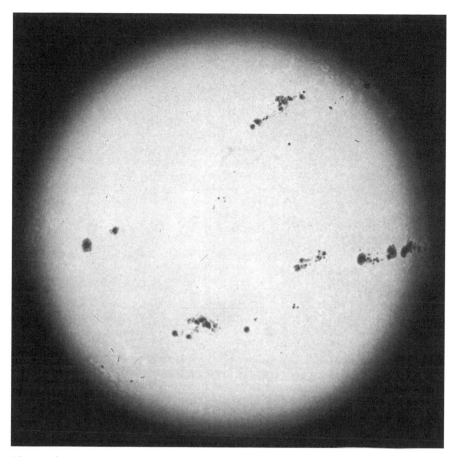

Photosphere of the Sun at sunspot maximum in December 1957. [Carnegie Observatories]

the white-hot photosphere, with temperatures of about 7,200°F (4,000°C). Its upper layers, however, are much hotter than the photosphere—about 18,000°F (10,000°C). This temperature causes hydrogen atoms to emit a red wavelength that is primarily responsible for the chromosphere's fiery color.[4] The chromosphere is a thin atmospheric layer, only about 1,600 miles (2,500 kilometers) thick, although there are no sharp boundaries above or below.

Seen under high magnification along the rim of the Sun, the chromosphere is not a smooth layer of gas. It looks like an erratic forest of thin, topless trees. These spiked features that compose the chromosphere are known as *spicules*. They are less than 400 miles (700 kilometers) in diameter but may tower thousands of miles high, reaching out of the chromosphere and into the corona. These spicules may be not only columns of rising gas but also the outlines of magnetic flux

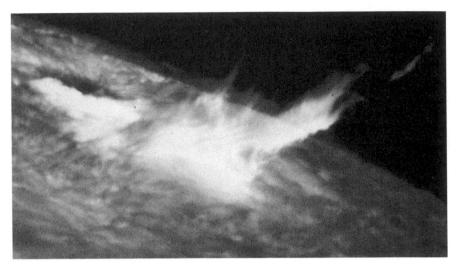

Solar flare. [National Optical Astronomy Observatories]

tubes that transfer magnetic fields from the photosphere to the corona.

It is in the photosphere and chromosphere that prominences and flares are rooted and stretch upward into the corona. They too are transient features of the Sun that vary with the rhythm of the sunspots below them. *Prominences* are the same temperature as the upper chromosphere and glow with the same red color of excited hydrogen. They are condensed clouds of solar gas, but much cooler and denser than the surrounding corona. The prominences are bent and twisted by local magnetic fields. Magnetic forces keep the gases in the prominences from all falling back to the surface, which would take only about 15 minutes if gravity were the only force acting. Most often these clouds, the prominences, do slowly rain material back toward the surface, but occasionally they erupt outward.

Flares are much stronger eruptions, also triggered by the magnetic activity of the Sun, that launch great torrents of mass and energy from the Sun at millions of miles an hour.

The chromosphere is also the birthplace of *solar tornadoes*, almost the size of Earth, whirling at speeds up to 300,000 miles per hour (500,000 kilometers per hour), spiraling up and through the corona, probably contributing fast-moving atomic fragments to the solar wind of particles flowing spaceward from the Sun.[5] How these solar tornadoes form and function is not certain, but, as with virtually every feature seen on the Sun's surface and in its atmosphere, the explanation almost certainly involves twisted magnetic fields.

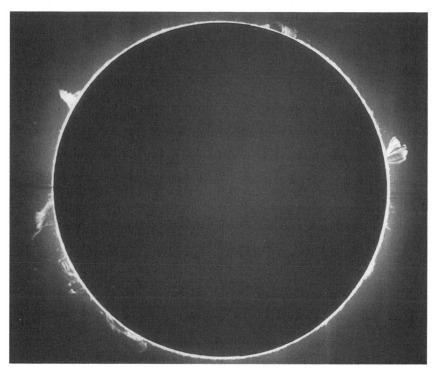

Prominences of different sizes and shapes silhouetted against the limb of the
Sun, December 9, 1929. [Carnegie Observatories]

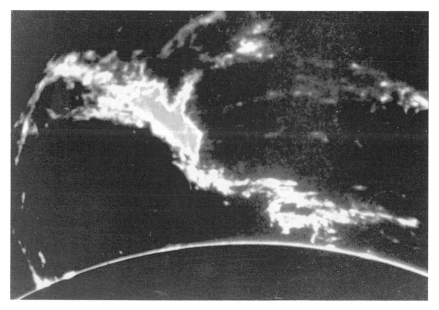

Eruptive prominence on the Sun. [NASA *Skylab*]

In the upper reaches of the chromosphere and extending outward into the corona is a realm called the *transition region*, so named because the temperature there suddenly climbs to over 1.8 million °F (1 million °C).

Why should the temperature of the *corona* ("crown"), the outer atmosphere of the Sun, be so high? It is much hotter than the photosphere and chromosphere, yet those layers are closer to the core where the Sun generates its energy. The answer again lies in the Sun's magnetic fields. The Sun not only behaves like a giant bar magnet (like the Earth with its magnetic poles), but the Sun is also pocked by many local magnetic regions with intensities greater than the Sun's polar magnetism. These surface sites of magnetic activity, the largest marked by sunspots, are induced by varying magnetic fields in the Sun's interior. These magnetic fields are borne to the surface by the columns of hot gases seen as granulations in the photosphere. The random motion of the rising and falling gases contort the lines of magnetic force. These magnetic fields stretch into the chromosphere and corona as loops and arches, as if they were invisible electric wires attached to terminals in the photosphere. The twisted and coiled magnetic fields induce electrical currents into the corona, which heat the gases there. But explaining exactly how this heating is accomplished in the near vacuum of the corona remains a challenge to solar astronomers.[6]

A Magnetic Personality

The gases rising from the surface of the Sun are all so hot that their atoms are missing some or all of their electrons. This plasma (ionized gas) expands as it rises from the surface of the Sun. As it expands, its density declines and these charged gases are more easily warped by the Sun's magnetic fields. The patterns of this magnetic control can

A loop prominence on the Sun, showing the magnetic lines of force, August 22, 1996. [ESA/NASA SOHO EIT]

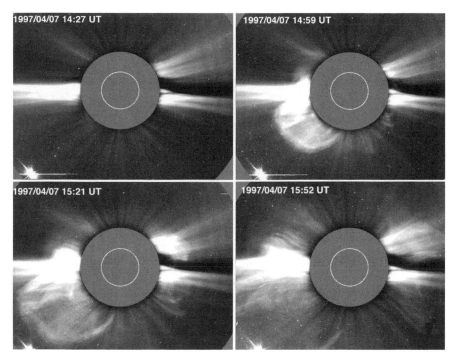

1997/04/07 14:27 UT

1997/04/07 14:59 UT

1997/04/07 15:21 UT

1997/04/07 15:52 UT

Coronal mass ejection (*to lower left of Sun*) observed by the SOHO spacecraft over 85 minutes, April 7, 1997. A coronagraph provides an artificial eclipse of the Sun. The circle on the occulting device shows the size of the Sun's disk. Subatomic particles and magnetic fields from this coronal mass ejection hit the Earth April 10-11. [ESA/NASA SOHO LASCO C2 coronagraph, courtesy Steele Hill]

be seen in the loops and arches of the prominences that extend high into the corona. Here and there a magnetic loop in the corona is stressed so greatly that it snaps and perhaps 10 billion tons of million-degree plasma is slung into space as a huge expanding bubble traveling as fast as 4.5 million miles per hour (2,000 kilometers per second). Such outpourings are called *coronal mass ejections.* Coronal mass ejections, eruptive prominences, and flares are probably different parts of the same phenomenon. They involve the expulsion of hot gases and twisted ropes of magnetic fields from the Sun, they often occur together, and all are more frequent at sunspot maximum.[8]

Coronal mass ejections enhance and create shock waves in the normal solar wind with charged particles (primarily protons and electrons) traveling 900,000 miles per hour (400 kilometers per second) or more that are escaping the Sun's gravitational hold. Coronal mass ejections also carry with them some of the coronal magnetic field, and distort the existing field in the solar wind. When the altered

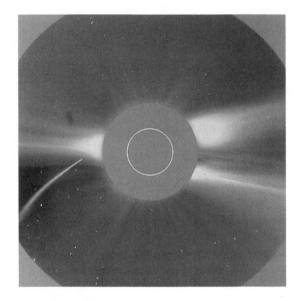

A Sun-grazing comet too near the Sun for its survival, December 23, 1996. [ESA/NASA SOHO LASCO C2 coronagraph, courtesy Steele Hill]

magnetic fields and these especially energetic subatomic particles strike the Earth, they create colorful displays of the northern and southern lights—the aurora. But solar windstorms can also damage electronic equipment on satellites in Earth orbit; disrupt telephone, radio, and television transmission; and overload electric power lines, causing blackouts (as one did for 9 hours to 6 million Canadians in Quebec on March 13, 1989). Scientists now think that coronal mass ejections, rather than flares, are the principal cause of the aurora and other geomagnetic events.

In visible light the corona shows graceful, delicate streamers and brushlike features.[9] However, in X ray pictures, which better capture the activity of high-temperature gases, the corona is a riot of ever-changing loops, plumes, eruptions, and contrasting light and dark regions. These dark regions of lesser activity are called *coronal holes*. It is primarily through these coronal holes, where magnetic fields are weaker, that the solar wind escapes into space. Slower and more variable solar wind flows from coronal streamers.

The coronal holes and streamers seen during a total eclipse mark the visible departure from the Sun of the solar wind. The Sun loses about 10 million tons of material a year in the solar wind. But such a loss is insignificant compared to the Sun's total mass and to the rate at which the Sun is converting mass into energy at its core.

The solar wind blows outward in all directions. The Earth is a small target 93 million miles (150 million kilometers) away, so only two out of every billion particles in the solar wind will reach the Earth, to cause mischief here. But they are enough. Scientists are studying "space weather" with the hope of predicting when the Earth will be struck by a stormy blast of enhanced solar wind.

To the eye, the white flame brushes of the corona can extend outward from the Sun 2 million miles (3 million kilometers) or more until they are so tenuous that they are no longer visible. But the corona is still measurable out to the Earth and beyond, in the form of the solar wind. The Earth orbits the Sun within the Sun's rarefied outer atmosphere.

We do not normally see the chromosphere or the corona. They are concealed from our view by the overwhelming glare of the photosphere, half a million times brighter than the corona. To study the Sun, early scientists had only its surface to rely on—just the photosphere and sunspots. Progress was slow.

We do not see the chromosphere or the corona unless something blocks the glare of the photosphere so that the faint atmosphere of the Sun is revealed. In the nineteenth century, astronomers discovered that the Moon was their scientific collaborator, obscuring the Sun's surface from time to time so that they might see a part of the Sun that had never been studied before.

In less than a century, total solar eclipses, and the scientific instruments and theories they helped to stimulate, revealed the composition of the Sun, the structure of the Sun's interior, and the wonder of how the Sun shines.

7

The First Eclipse Expeditions

I did not expect, from any of the accounts of preceding eclipses that I had read, to witness so magnificent an exhibition as that which took place.

—Francis Baily (1842)

An Unlikely Beginning

Francis Baily, the man who might be said to have founded the field of solar physics, received only an elementary education, was not trained in science, and did not get around to astronomy until the age of 37. Like his father, a banker, he entered the commercial world as an apprentice when he was 14. But adventure called. When his seven years of apprenticeship expired, he sailed for the New World and spent the next two years, 1796–1797, exploring unsettled parts of North America, narrowly escaping from a shipwreck, flatboating down the Ohio and Mississippi Rivers from Pittsburgh to New Orleans, and then hiking nearly 2,000 miles back to New York through territory inhabited mostly by Indians. He liked the United States so well that he planned to marry and become a citizen, but he finally abandoned those plans and returned home in 1798.

Back in England, he began efforts to mount an expedition to explore the Niger River in Africa. He could not raise enough money, however, so he became a stockbroker. To dedication and enthusiasm he quickly added a reputation for intelligence and integrity, and he made a fortune. He exposed stock-exchange fraud and helped clean it up. He published a succession of explanations of life insurance methods and comparisons of insurance companies, which became wildly

Francis Baily. [Royal
Astronomical Society]

popular. He also published a chart of world history that was equally
popular, confirming the nickname given to him in his apprentice
days: the Philosopher of Newbury (his birthplace).

His first astronomical paper (1811) tried to identify the solar
eclipse allegedly predicted by Thales. In 1818, he called attention to
an annular eclipse of the Sun coming in 1820, and he observed it
from southeastern England. That same year, he became one of the
founders of the Astronomical Society of London, later the Royal
Astronomical Society.

In 1825, he retired from the stock market to devote all his time to
his new profession. He was 51 years old. His revisions of a series of
old star catalogs were considered so valuable that the Royal Astronomical Society twice awarded him its Gold Medal and four times
elected him president. Although he was not renowned as an observer,
he had an abiding fascination with eclipses, a good eye for detail, and
the ability to express what he saw.

Thus it was in 1836 that a few words from Francis Baily sparked
the immediate, intense, and unending study of the physical proper-
ties of the Sun that had been generally ignored or discounted until
then. He traveled to an annular eclipse of the Sun in southern Scotland and watched on May 15, 1836, as mountains at the Moon's limb
occulted the face of the Sun but allowed sunlight to pour through the

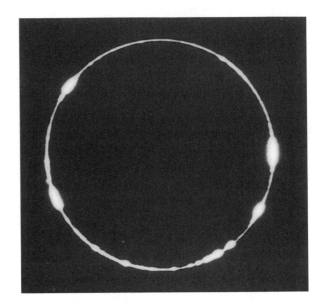

Baily's Beads during the annular eclipse of April 28, 1930. [Lick Observatory]

valleys between them so that the ring of sunlight around the rim of the Moon was broken up into "a row of lucid points, like a string of bright beads."[1] With those words, Baily generated fervor for solar physics and founded the industry of eclipse chasing.

The Surprise of Totality

At the next accessible eclipse, July 8, 1842, a high percentage of the astronomers of Europe migrated to southern France and northern Italy to see "Baily's Beads." Baily, now 68 years old, went too.

This was not an annular eclipse, as Baily had seen twice before. It was total. No European astronomer then alive had ever seen a total eclipse.

Baily set up his telescope at an open window in a building at the university in Pavia, Italy: "[A]ll I wanted was to be left *alone* during the whole time of the eclipse, being fully persuaded that nothing is so injurious to the making of accurate observations as the intrusion of unnecessary company."[2]

Again Baily's Beads were visible—to Baily at least. George Airy, England's astronomer royal, observing from Turin, Italy, did not see them. Baily was just jotting down the time of appearance and duration of the beads

. . . when I was astounded by a tremendous burst of applause from the streets below, and at the *same moment* was electrified at the sight of one of the most brilliant and splendid phenomena that can well be imagined. For, at that instant the dark body of the moon was *suddenly* surrounded with a *corona*, or kind of bright *glory*, similar in shape and relative magnitude to that which painters draw round the heads of saints, and which by the French is designated an *auréole*.

Baily was not the first to use the word *corona* to designate the glowing outer atmosphere of the Sun visible during a total eclipse, but his striking description of it caught everyone's attention as it never had before and forever united the word with the phenomenon.

[W]hen the total obscuration took place, which was *instantaneous*, there was an universal shout from every observer . . . I had indeed anticipated the appearance of a luminous circle round the moon during the time of total obscurity: but I did not expect, from any of the accounts of preceding eclipses that I had read, to witness so magnificent an exhibition as that which took place. . . . It riveted my attention so effectually that I quite lost sight of the string of *beads*, which however were not completely closed when this phenomenon first appeared.

There was so much to see, said Baily, that in future eclipses, each observer should be assigned a single observing task.

Splendid and astonishing, however, as this remarkable phenomenon really was, and although it could not fail to call forth the admiration

Drawings of corona and prominences. *Left*: July 8, 1842; *right*: July 28, 1851. [François Arago: *Popular Astronomy*, edited and translated by W. H. Smyth and Robert Grant]

and applause of every beholder, yet I must confess that there was at the same time something in its singular and wonderful appearance that was appalling: and I can readily imagine that uncivilised nations may occasionally have become alarmed and terrified at such an object, more especially in times when the true cause of the occurrence may have been but faintly understood, and the phenomenon itself wholly unexpected.

It was the last eclipse Francis Baily was to see. Two years later he died. Of him, historian Agnes M. Clerke wrote: "He was gentle as well as just; he loved and sought truth; he inspired in an equal degree respect and affection. . . . Few men have left behind them so enviable a reputation."[3]

Before Baily

Never again would a total eclipse over an inhabited landmass go unattended by professional and amateur astronomers, even when the observers had to travel to remote sites halfway around the world. The corona, Baily's Beads, prominences, shadow bands—all had been seen before, many times, throughout the world. But now they commanded attention and explanation.

The first written record of the corona may be Chinese characters inscribed on oracle bones from about 1307 B.C. that say "three flames ate up the Sun, and a great star was visible."[4] Or were these flames prominences instead? The first unequivocal description of the corona comes from a chronicler more than 2,000 years later, observing from Constantinople the eclipse of December 22, 968 A.D.

The great astronomer Johannes Kepler did not see a total eclipse himself, but from the reports he read, he concluded that the corona must be material around the Sun and not the Moon. Giacomo Filippo Maraldi, an Italian-born French astronomer, provided evidence that the corona is part of the Sun because the Moon traverses the corona during a solar eclipse; the corona does not move with the Moon but stays fixed around the Sun.

The Spanish astronomer José Joaquin de Ferrer, well ahead of his time, traveled to the New World to observe total eclipses in Cuba in 1803 and Kinderhook, New York, in 1806. He was probably the first to use the word "corona" to describe the glow of the outer atmosphere of the Sun seen during a total eclipse.

The disk [of the Moon] had round it a ring of illuminated atmosphere, which was of a pearl colour . . . From the extremity of the ring, many luminous rays were projected to more than 3 degrees distance.—The lunar disk was ill defined, very dark, forming a contrast with the luminous corona . . .[5]

Ferrer quite correctly attributed the corona to the Sun. If this glow belonged to the Moon, he calculated, the lunar atmosphere would extend upward 348 miles (560 kilometers)—50 times more extensive than the atmosphere of Earth. Thus, he concluded, the corona "must without any doubt belong to the Sun." Baily, too, at the 1842 eclipse, attributed the corona to the Sun.

The first sure report of solar prominences came from Julius Firmicus Maternus in Sicily, who noticed them during the annular eclipse of July 17, 334 A.D. Edmond Halley, the great English astronomer, saw them clearly as bright red protrusions during the total eclipse of May 3, 1715.

Baily had pointed the way. But what was the significance of the corona and the prominences? New techniques—photography and spectroscopy—were emerging with the capability to record and explore these features of the Sun. In photography and spectroscopy, observational astronomy gained its two most powerful tools to augment the telescope.

The stage was set for the eclipse of July 28, 1851. The astronomical world gathered along its path of totality through Scandinavia and Russia. They solved one mystery and uncovered another.

The Debut of Photography

The first successful photograph of the Sun in total eclipse was a daguerreotype taken on July 28, 1851, by a professional photographer named Berkowski, assigned to the task by August Ludwig Busch, director of the observatory in Königsberg, Prussia.[6] The inner corona and prominences are clearly visible. Yet, for now, photography was just a curiosity, a promising experiment. Hard-core science was still carried out by visual observations made through telescopes.

At the 1851 eclipse, two teams of astronomers provided proof of what most astronomers suspected: that prominences were part of the Sun. Robert Grant and William Swan from the United Kingdom and Karl Ludwig von Littrow from Austria documented how the eastward motion of the Moon across the face of the Sun covered prominences along the east rim of the Sun as totality began and then uncovered prominences along the west rim as totality ended, demonstrating that the prominences belonged to the Sun and that the Moon was only passing in front of them.

Often in science, observations that lead to the solution of one problem simultaneously reveal a fascinating new problem. As George B. Airy, England's astronomer royal, was observing this eclipse, he noticed a jagged edge to the solar atmosphere just above the edge of the Moon. He called it the *sierra*, thinking that he might be looking at mountains on the Sun.

Airy thus became the first to call attention to the chromosphere, the lowest level of the transparent solar atmosphere immediately above the opaque photosphere that creates the appearance of a surface for the Sun. The jaggedness that Airy observed was actually innumerable small jets of rising gas called spicules.

By 1860, daguerreotypes were obsolete, superseded by the faster wet-plate (collodion) photographic process. Before photography, astronomers could only describe or sketch what they saw. The early photographic emulsions were not as sensitive to detail as desired, but they were more objective than the human eye.

On July 18, 1860, a total eclipse was visible from Europe and found astronomers waiting with improved cameras. Prominences remained a matter of high priority and they encountered the resourcefulness of British astronomer Warren De La Rue and Italian astronomer and Jesuit priest Angelo Secchi.

De La Rue's well-to-do family provided him with a solid education, after which he entered his father's printing business. There he demonstrated a great affinity for machinery. He could make any instrument or device run better. He was also a fine draftsman, and it

First photograph of the Sun in total eclipse, July 28, 1851. [Courtesy of Dorrit Hoffleit]

Drawing by Lilian Martin-Leake from a telescopic view of the chromosphere and the corona, May 28, 1900, showing red spicules in the chromosphere that George Airy had thought were mountains. [Annie S. D. Maunder and E. Walter Maunder: *The Heavens and Their Story*]

was this talent that lured him into astronomy. He could produce drawings of the planets, Moon, and Sun that were better than those of astronomers. But no sooner had he produced his first excellent drawings than he found out about a new invention called photography. He never saw a mechanical device that did not fascinate him, so he was soon making improvements in cameras, inventing specialized cameras for solar photography, and photographing the Moon and Sun stereographically so that lunar features appeared in relief and sunspots revealed themselves as depressions, not mountains, in the photosphere.

Observing the 1860 eclipse 250 miles (400 kilometers) away from De La Rue was Angelo Secchi. He too had a remarkable aptitude for inventing and coaxing instruments. Secchi was from a poor family and received his education through the Catholic Church in the demanding Jesuit tradition. From the beginning, he showed brilliance in mathematics and astronomy. When the Jesuits were expelled from Italy by a liberal, anticlerical government in 1848, Secchi spent a year in the United States as assistant to the director of the Georgetown University observatory. When the ban against Jesuits was lifted in 1849, he returned to Italy to become director of the Pon-

Angelo Secchi. [Mabel Loomis Todd: *Total Eclipses of the Sun*]

tifical (now Vatican) Observatory of the Collegio Romano (or Gregorian University). He transformed it into a modern, well-equipped center for research in the new field of astrophysics. Secchi was one of the pioneers in applying spectroscopy to astronomy. He surveyed more than 4,000 stars and realized that stellar spectra could all be grouped into a handful of classifications.

De La Rue and Secchi ambushed the 1860 eclipse with improved cameras using the wet-plate process, which greatly reduced exposure time and thereby increased the clarity with which objects in motion could be seen. They captured the prominences and compared them. The prominences looked the same from the photographers' widely separated sites, so Secchi and De La Rue could conclude that they were indeed part of the Sun. If they had been features on the Moon, so much closer than the Sun, the difference in viewing angles (parallax) from Secchi's and De La Rue's separate sites would have given them a different appearance.

The Debut of Spectroscopy

On August 18, 1868, a great eclipse touched down near the Red Sea and swept across India and Malaysia. Once again, the international scientific community had assembled.

From the United Kingdom to the path of the eclipse had come, among others, James Francis Tennant and John Herschel (son of John F. W. Herschel, grandson of William Herschel, both renowned astronomers). Norman Pogson, born in England, represented India and the observatory he directed there. The French delegation included Georges Rayet and Jules Janssen. Each of them carried a new weapon just added to the scientific arsenal for prying secrets from the Sun during an eclipse. That tool was a spectroscope. It would prove as indispensable to eclipse studies of the structure of the Sun as it was rapidly demonstrating itself to be in other realms of astronomy and in all other physical and biological sciences. By passing the light of the corona or the prominences through a prism, it could be broken down into a spectrum of lines and colors. From this spectrum, a scientist could identify the chemical elements present and even the temperature and density of its source.

As the eclipse sped along its course, spectroscopes pointed upward toward the prominences. They showed that the prominences emitted bright lines, and most of these were quickly identified with hydrogen. More and more it seemed that the Sun must be composed primarily of gas and that hydrogen must be a major constituent.[8] The spectroscopists, properly pleased with their results, packed their equipment and headed home.

All but one. His name was Jules Janssen and the brightness of the prominences and the strength of their spectral lines had given him an idea. He wanted to look for them again when the eclipse was over, when the Moon was not blocking the intense glare of the Sun from view. Might it be possible for him to see the prominences and their spectrum in broad daylight?

The weather was cloudy the rest of that day. He would have to wait until tomorrow.

That Extra Step

Pierre Jules César Janssen was 12 years old when Baily called attention to the beads visible during the annular eclipse of 1836. A childhood accident had left Janssen lame and he never attended elementary or high school. His family was cultured, but his father was a struggling musician, so Jules had to go to work at a young age.

While employed at a bank, he earned his college degree in 1849. He then went on to gain a certificate as a science teacher and served as a substitute teacher at a high school. In 1857, he traveled to Peru as part of a government team to determine the position of the magnetic equator. There he became severely ill with dysentery and was

Jules Janssen (1824–1907)

Two years after his breakthrough at the 1868 eclipse, Jules Janssen again planned to apply spectral analysis to a solar eclipse, this one on December 22, 1870, in Algeria. When it came time for departure, however, the Franco-Prussian War was in progress and Paris was under siege. Colleagues in Britain had obtained from the Prussian prime minister safe passage for Janssen from Paris, but Janssen wouldn't accept favors from his country's enemies. He had a different plan: "France should not abdicate and renounce taking part in the observation of this important phenomenon. . . . [A]n observer would be able, at an opportune moment, to head toward Algeria by the aerial route . . . "*

Although he had never been in a balloon before, on December 2, with a sailor as an assistant and himself as pilot, he ascended from Paris and headed west. Despite violent winds, he landed safely near the Atlantic coast. He reached Algeria in time—only to have the eclipse clouded out. From that experience, however, Janssen designed an aeronautical compass and ground speed indicator and prophesied methods of air travel that would "take continents, seas, and oceans in their stride." In 1898 and subsequently, he used balloons to study meteor showers from above the clouds, pioneering high-altitude astronomy and foreseeing the advantages of space observations.

Janssen was also a leader in astrophotography. "[T]he photographic plate is the retina of the scientist," he wrote. It was for spectroscopy, however, that Janssen was most renowned. His methods opened up the Sun's atmosphere to continuous study. The French government tried to find him an observatory position, but the director of the Paris Observatory did not want him. So Janssen was allowed to pick a site near Paris for a new observatory dedicated to astrophysics. He chose Meudon and directed the observatory from its founding in 1876 until his death in 1907.

"There are very few difficulties that cannot be surmounted by a firm will and a sufficiently thorough preparation," he wrote. But he was too modest in his self-assessment. To everything he investigated, he brought imagination and insight. Jules Janssen was one of the most creative scientists of any era.

---

*Quotations are from the sketch of Janssen by Jacques R. Lévy in the *Dictionary of Scientific Biography*.

---

sent home. At age 33, he seemed destined for a quiet life in teaching, if he could get a job. He became the tutor for a wealthy family in central France. At their steel mills, he noticed that the eye could watch molten metal without fatigue or injury, while the skin had to be protected from the heat. He wrote a careful study of how the eye protects

Jules Janssen. [Mary Lea Shane Archives of the Lick Observatory]

itself against heat radiation, which earned him his doctorate in 1860.

The glow of molten metal had led him to spectroscopic analysis, which he then applied to the Sun and, in 1859, identified several Earth elements present in the Sun. He moved to Paris in 1862 to dedicate himself to solar physics and scientific instrument-making. His work had already made him a leader in solar spectroscopy. He used the changing spectrum of the Sun in its daily journey from horizon to horizon to separate spectral lines caused by the Earth's atmosphere from those originating in the Sun and to demonstrate the composition and density of the Earth's atmosphere through which the sunlight passed. Janssen then applied spectroscopy to the other planets and, in 1867, discovered water in the atmosphere of Mars.

He traveled to Guntur, India, for the total eclipse of August 18, 1868, to use his spectroscope on solar prominences. The great contemporary English spectroscopist Norman Lockyer spoke ringingly of that pivotal moment: "Janssen—a spectroscopist second to none . . . was so struck with the brightness of the prominences rendered visible by the eclipse that, as the sun lit up the scene, and the prominences disappeared, he exclaimed, '*Je reverrai ces lignes la!*'[9] [I will see those lines again!]" The next morning he succeeded. He had found a way to study the atmosphere of the Sun without waiting for a total eclipse, traveling halfway around the world to see it, and hoping for good weather at the immutable moment.

For two weeks Janssen continued to map solar prominences by this technique and continued to perfect it on his circuitous way home, with a stop in the Himalayas to observe at high altitude. He proved that prominences change considerably from one day to the next.

A standard spectroscope breaks down the light of a glowing object into the characteristic colors of its spectrum. Janssen modified the spectroscope by blocking unwanted colors so that the observer could view an object in the light of one spectral line at a time. He had invented the spectrohelioscope. The Sun could now be analyzed in detail on a daily basis.

A month after the eclipse, on his way home to France, Janssen wrote up his findings and sent them to the Academy of Sciences in Paris. His paper arrived a few minutes after one from England that reported precisely the same discovery.

Coincidence

Joseph Norman Lockyer came from a well-to-do family with scientific interests. He received a classical education, traveled in Europe, and then entered civil service. So wide were his interests that he wrote on everything from the construction dates and astronomical purposes of Egyptian pyramids and temples to Tennyson to the rules of golf. "The more one has to do, the more one does" was his motto.[10]

When Gustav Kirchhoff and Robert Bunsen showed in 1859 how spectroscopy could be used to determine the chemical composition of objects in space, Lockyer saw the discovery as a key to what had seemed the locked door of the universe. French philosopher Auguste Comte had confidently asserted only 24 years earlier that never, by any means, would we be able to study the chemical composition of celestial bodies, and every notion of the true mean temperature of the stars would always be concealed from us.[11] It was clear that Comte was wrong. Lockyer bought a spectroscope, attached it to his 6¼-inch (16-centimeter) refracting telescope, and began his observations.

Although he had never seen a solar eclipse, it occurred to him that, since prominences were probably clouds of hot gas, he should be able to use a spectroscope to analyze prominences without waiting for an eclipse. This idea struck Lockyer two years before the 1868 eclipse that inspired Janssen to the same realization. Lockyer tried the experiment in 1867 but found his spectroscope inadequate to the task. So he ordered a new spectroscope to his specifications. Because of construction delays, however, it did not arrive until October 16, 1868, two months after the eclipse that Janssen saw in India. Lockyer

Medallion created by the Academy of Sciences in Paris to honor Janssen and Lockyer for their independent discovery of how to observe prominences without waiting for an eclipse. The front of the medal shows the heads of the two scientists. The reverse shows the Sun god Apollo pointing to prominences on the Sun.

rapidly and excitedly calibrated his new instrument and on October 20, 1868, he trained it on the rim of the Sun and recorded bright lines typical of hot gases under very little pressure. He wrote up his findings and sent them to the Academy of Sciences in Paris for presentation by his friend Warren De La Rue.

Just minutes before De La Rue was to speak, Janssen's letter arrived and both papers were read at the same session of the Academy to great acclaim for both scientists. A special medal was created to honor them. It showed the heads of Janssen and Lockyer side by side.[12]

A New Element

Lockyer continued to examine the spectrum of the gases at the rim of the Sun. He recognized that the lower atmosphere of the Sun, what Airy had called the sierra, was decidedly reddish in color, so he named it the *chromosphere*, and it has been known by that name ever since.

Lockyer was not done yet. In examining the spectrum of the prominences, he noticed a yellow line that he could not identify. It did not seem to belong to any element known on Earth. So he announced the existence of a new element and proposed the name *helium* for it, because it had been found in the Sun—*helios* in Greek. Most scientists rejected the idea of a new element, suggesting that this line was produced by a known element under unusual physical conditions. But Lockyer clung tenaciously to his interpretation. Finally,

J. Norman Lockyer (1836–1920)

In 1869, the year after he independently showed how the atmosphere of the Sun could be analyzed without the benefit of an eclipse and discovered the element helium, Lockyer founded the scientific journal *Nature*. He edited it for 50 years, until just before his death, keeping it alive through many crises.

While the French government was establishing a special observatory for Janssen, the British government likewise recognized the importance of Lockyer's contribution and set about creating a solar physics observatory for him. For its opening in 1875, Lockyer collected old and modern instruments and placed them on display. The display became permanent and grew. Lockyer had founded London's world famous Science Museum.

Lockyer was not shy in interpreting his findings to form startling theories. Often he was wrong, but always he provided useful data and often there was a nucleus of truth in his grand speculations. He thought that all atoms shared certain spectral lines and were therefore made of smaller common constituents. He was wrong about the spectra, but right about the composition of atoms.

He offered dates for the construction of ancient Egyptian temples based on their alignments with the rising and setting positions of the Sun and certain stars. His dates were wrong, but he was right that many of the temples had astronomical orientations and his work helped to establish the field of archaeoastronomy.

When he died, a colleague wrote of him: "Lockyer's mind had the restless character of those to whom every difficulty is a fresh inspiration. His enthusiasm never failed him, despite repeated disappointments and opposition."*

*Alfred Fowler: "Sir Norman Lockyer, K.C.B., 1836-1920," *Proceedings of the Royal Society of London*, series A, volume 104, 1923, pages i-xiv.

---

in 1895, William Ramsay found trapped in radioactive rocks on Earth an unknown gas that exhibited the mysterious spectral line that Lockyer had discovered on the Sun. Helium was an element. Lockyer had been right. In 1869, just after he discovered helium, Lockyer had urged: "[L]et us . . . go on quietly deciphering one by one the letters of this strange hieroglyphic language which the spectroscope has revealed to us—a language written in fire on that grand orb which to us earth-dwellers is the fountain of light and heat, and even of life itself."[13]

The total eclipse of 1868 had raised the strong possibility of the existence of a new element—and, for the first time, one discovered

J. Norman Lockyer in 1895, the year helium, the element he discovered on the Sun, was finally found to exist on Earth as well.

not on Earth but in the heavens. One year later, on August 7, 1869, the United States lay on the path of a total eclipse. Two American astronomers, Charles A. Young and William Harkness, working separately, observed the event with spectroscopes. Each noticed a green line in the spectrum of the corona that defied identification with known elements. This suspected new element was called *coronium*.

In 1895, Lockyer's helium was identified in rocks on Earth, but coronium remained a spectral presence seen only on the Sun during total eclipses. As time passed, more elements were discovered on Earth until the periodic table of stable chemical elements was nearly complete. There was no room left for coronium to be an element. What could it be?

Might it be an already known element under such unusual conditions that it emitted a spectrum never before seen in a laboratory? Walter Grotrian of Germany in 1939 pointed the way and Bengt Edlén of Sweden in 1941 identified the green line of coronium as the element iron with 13 electrons missing—a "gravely mutilated state."[14] To ionize iron so greatly, the temperature of the corona had to be about 2 million °F (1 million °C) and its density had to be less than a laboratory vacuum. Because the conditions necessary for the production of such lines cannot be achieved in a laboratory, they are known as forbidden lines.

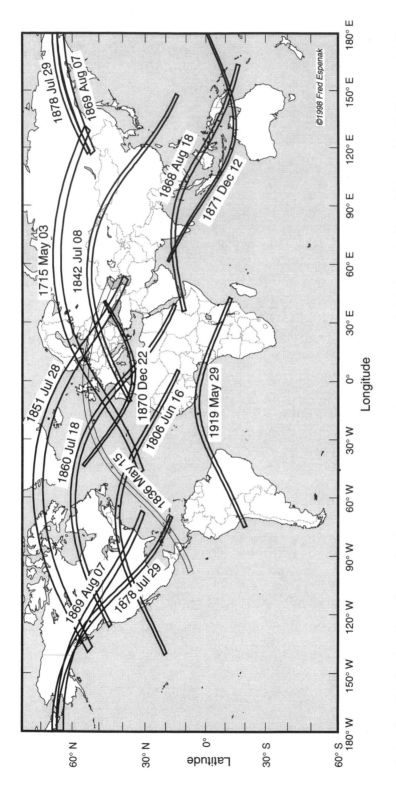

Paths of totality for the solar eclipses of 1715, 1806, 1836, 1842, 1851, 1860, 1868, 1869, 1870, 1871, 1878, and 1919. [Map and eclipse calculations by Fred Espenak]

Charles A. Young. [Mary Lea
Shane Archives of the Lick
Observatory]

The Reversing Layer

Astronomy was a family tradition for Charles Augustus Young. His
maternal grandfather and his father had been professors of astronomy at Dartmouth College. Charles entered Dartmouth at age 14 and
four years later graduated first in his class. He immediately began
teaching—classics!—at an elite prep school, and commenced studies
at a seminary to become a missionary. But in 1856 he changed his
plans and became professor of astronomy at Western Reserve College,
with a break in his duties to serve in the Civil War. He returned to
Dartmouth in 1866 at the age of 32 as professor of astronomy, holding the same chair his father and grandfather held before him. There
he pioneered in spectroscopy, especially as applied to the Sun.

At the eclipse of December 22, 1870, which he observed at Jerez,
Spain, Young noticed that the dark lines in the Sun's spectrum
become bright lines for a few seconds at the beginning and end of
totality. He had discovered the *reversing layer*, the lowest 600 miles
(1,000 kilometers) of the chromosphere, which is cooler than the
photosphere and thus absorbs radiation of specific wavelengths, producing the ordinary dark-line spectrum of the Sun. However, when
the Moon blocks the photosphere from view and the reversing layer

What Eclipses at Jupiter Taught Us

by Carl Littmann

As the moons in the solar system revolve around their planets, they too create and undergo periodic solar and lunar eclipses. These events allowed the Danish astronomer Ole Römer in 1676 to prove, contrary to prevailing opinion, that light travels at a finite speed. He even succeeded in making the first good estimate of the speed of light. Such luminaries as Aristotle, Kepler, and Descartes had been certain that the speed of light was infinite. Gian Domenico Cassini, director of the Paris Observatory where Römer made his discovery, refused to believe the results, and Römer's triumph was not fully appreciated for half a century.

(left) Solar eclipse on Jupiter. Black dot at the left is the shadow of Ganymede, Jupiter's largest moon. At the far right is Io, Jupiter's innermost large satellite. [NASA/Jet Propulsion Laboratory *Voyager 2*]

(right) Solar eclipse on Saturn caused by its ring system. At the top are moons Tethys and Dione. Tethys' shadow on Saturn can be seen at the lower left, just above the shadow of the rings. [NASA/Jet Propulsion Laboratory *Voyager 1*]

of the chromosphere can be seen momentarily before it too is eclipsed, the bright-line spectrum of its glowing gases briefly flashes into view. Here at last was the layer responsible for the dark-line spectrum of the Sun seen on ordinary days. A long-missing piece in the puzzle of the structure, composition, and density of the solar atmosphere was fitted into place.

In 1877, Young was lured away from Dartmouth by the offer of more equipment and research time at the College of New Jersey, now Princeton University. He was not only a great researcher but also a revered teacher and acclaimed writer. His textbooks were the standard of his day.

Saturn's shadow eclipses its rings. [NASA/Jet Propulsion Laboratory *Voyager 1*]

In observations of Io, innermost of Jupiter's four large moons, Römer noticed discrepancies between the observed times of its disappearance into the shadow of the planet and the calculated times for these events. He correctly explained that these discrepancies were due to the travel time of light between Jupiter and the Earth. When the Earth is approaching Jupiter, the interval between satellite eclipses is shorter because the distance light must travel is decreasing. When the Earth is moving farther from Jupiter, the interval between eclipses lengthens because the distance light must travel is increasing. Römer determined that light from Io took about 22 minutes longer to reach the Earth when the Earth was farthest from Jupiter than when it was closest. Thus, light required about 22 minutes to cross the orbit of Earth. The diameter of the Earth's orbit was not known at the time. Modern measurements show that light actually requires about $16\frac{2}{3}$ minutes to make this journey.

When Römer returned to Denmark, the king gave him a succession of appointments, including master of the mint, chief judge of Copenhagen, chief tax assessor (everyone said he was fair!), mayor and police chief of Copenhagen, senator, and head of the state council of the realm—all this and more while he served as director of the Copenhagen observatory and astronomer royal of Denmark. He discharged all his duties with distinction.

The Legacy of Eclipses

Throughout the final three decades of the nineteenth century, Janssen, Lockyer, and Young led expeditions to the major total eclipses and, weather permitting, always contributed useful data and often new discoveries.

Many scientists had noticed that the corona changes its appearance from one eclipse to the next. But it was Jules Janssen who first spotted a pattern to those variations. He compared the coronas of the 1871 and 1878 eclipses and concluded that the shape of the corona varies according to the sunspot cycle. In 1871, the Sun was near

Bolivia, November 12, 1966; dot to right is Mercury

Mexico, March 7, 1970

India, February 16, 1980

Siberia, July 31, 1981

Java, June 11, 1983

Philippines, March 18, 1988

The shape of the corona changes with sunspot activity. Near sunspot minimum, the corona is elongated at the Sun's equator. Near maximum, the corona is more symmetrical. These photographs were taken with a radial density filter developed by Gordon Newkirk that captures faint detail in the outer corona without overexposing the inner corona—similar to what the eye sees. The corona never appears the same twice. [High Altitude Observatory/National Center for Atmospheric Research]

sunspot maximum and the corona was round. In 1878, near sunspot minimum, the corona was more concentrated at the Sun's equator.

As the twentieth century began, solar eclipses were still the principal means of gathering information about the workings of the Sun. Every total eclipse over land was attended by scientists willing to travel great distances, endure hostile climates, and risk complete failure because of clouds for a few minutes' view of the corona—vital for the systematic study of the Sun launched more than half a century earlier by Francis Baily in his report on the annular eclipse of 1836.

One could see the Sun best when it was obscured.

8

The Eclipse That Made Einstein Famous

Oh leave the Wise our measures to collate
One thing at least is certain, light has weight
One thing is certain, and the rest debate—
Light-rays, when near the Sun, do not go straight.
                                        —Arthur S. Eddington (1920)

Of all the lessons that scientists learned from eclipses, the most profound and momentous was the confirmation of Einstein's general theory of relativity by the eclipse of May 29, 1919.

The Birth of Relativity

In 1905, an obscure Swiss patent examiner third class named Albert Einstein published three articles in the same issue of the leading German scientific journal *Annalen der Physik* that utterly changed the course of physics. One proved the existence of atoms. The second laid the cornerstone for quantum mechanics. The third was a revolutionary view of space and time known as the special theory of relativity. It required only high-school algebra, yet its implications were so profound that this theory baffled many of the leading scientists of the day.

In the decade that followed the publication of the special theory of relativity, Einstein labored mightily to expand his concept to accelerated systems. In 1907, he formulated his principle of equivalence: there is no way for a participant to distinguish between a gravitational system and an accelerated system. For an observer in a closed compartment, does a ball fall to the floor by gravity because the compartment is resting on a planet or does the ball fall because the compartment is accelerating toward the ball? In both cases, the ball falls to the floor. In both cases, the observer feels weight. It is impossible to tell whether that weight is from gravity or acceleration.

Einstein then realized from his principle of equivalence that relativity required gravity to bend light rays, much as it bends the paths of particles. In a closed accelerating compartment, a light on one wall is aimed directly at the opposite wall, across the line of motion. The beam travels at a finite speed, so in the time it takes to traverse the compartment, the opposite wall has moved upward. For an observer in the compartment, the light beam has struck the opposite wall below where it was aimed. The observer concludes that light has been bent.

What about that same experiment performed in a compartment at rest on a planet? According to Einstein's principle of equivalence, there can be no difference between phenomena measured in the two compartments. Therefore, for the observer in the gravitational environment, light must be bent by gravity. But the effect is very small. It takes a lot of mass to bend light enough to be measured. In 1911,

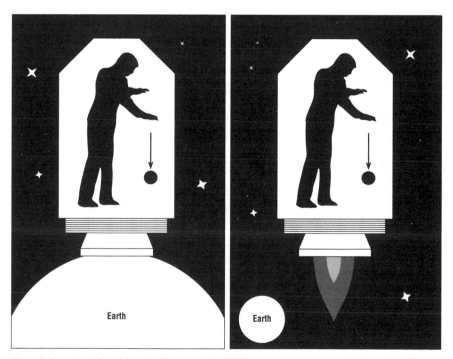

Einstein's principle of equivalence. A ball falls in a compartment. Does it fall by gravity because the compartment is resting on a planet or does it fall because the compartment is accelerating upward? Einstein realized that an observer in the compartment could not distinguish whether gravity or acceleration caused the ball to fall.

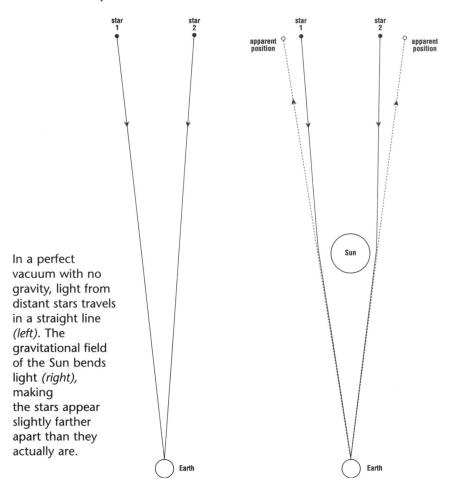

In a perfect vacuum with no gravity, light from distant stars travels in a straight line *(left)*. The gravitational field of the Sun bends light *(right),* making the stars appear slightly farther apart than they actually are.

Einstein realized that this peculiar idea might be tested during a total solar eclipse.

The Sun is sufficiently massive that light from distant stars passing close to its surface would be deflected just enough to be measurable. This bending of the light from such stars causes them to appear displaced slightly outward from the Sun. The displacement could be recognized by comparing a photograph of the star field around the Sun with a photograph of the same star field when the Sun was not present.

When the Sun is visible, however, the stars are much too faint to be seen. If only the Sun's brightness could be shut off for just a few minutes—as happens during a total solar eclipse! Using his special theory of relativity, Einstein initially predicted that starlight just grazing the Sun would be bent 0.87 arc seconds.[1]

The Search for Proof

A talented young scientist named Erwin Freundlich was the first to attempt to test this aspect of relativity.[2] Fascinated by the theory, he obtained solar eclipse photographs from observatories around the world that might show the displacement of stars. But photographs of previous eclipses were not adequate for the purpose.[3] The theory would have to be tested at a future eclipse, such as the one that would occur in southern Russia on August 21, 1914.[4] Freundlich was an assistant at the Royal Observatory in Berlin and tried to interest his colleagues in mounting a scientific expedition to test the theory. His superiors, however, were uninterested. Freundlich was allowed to go if he took unpaid leave and raised his own money. With youthful enthusiasm, he made his plans and informed Einstein of his intentions.

Einstein had just moved from Switzerland to Berlin to take a distinguished position created for him at the Kaiser Wilhelm Institute, the most prestigious research center in the world's science capital. Even among the giants there, Einstein stood out. Physicist Rudolf Ladenburg recalled, "There were two kinds of physicists in Berlin: on the one hand was Einstein, and on the other all the rest."

It was at this time, the spring of 1914, that Einstein was becoming increasingly withdrawn and oblivious to conventions of social behavior. He was confident in his new general theory of relativity but was deeply engrossed in its final formulation and its implications. Freundlich's wife Käte told of inviting Einstein to dinner one evening. At the conclusion of the meal, as the two scientists talked, Einstein suddenly pushed back his plate, took out his pen, and began to cover their prized tablecloth with equations. Years later, Mrs. Freundlich lamented, "Had I had kept it unwashed as my husband told me, it would be worth a fortune."[5]

That summer Freundlich took his scientific equipment and headed for the Crimea and the eclipse. On August 1 Germany declared war on Russia, commencing World War I. Freundlich was a German behind Russian lines. He and his team members were arrested and their equipment impounded. Within a month, Freundlich and crew were exchanged for high-ranking Russian officers, but they had missed the eclipse.[6]

Einstein deplored the war and German militarism, an attitude that drew the ire of many of his colleagues in Berlin. He ignored the hostility and concentrated all his energies on his research. In 1915 Einstein announced the completion of his general theory of relativity, a radically new theory of gravitation. It is perhaps the most prodigious work ever accomplished by a human being. Not only were its implications profound, but the mathematics needed to understand it were formidable. "Compared with [the general theory of relativity]," said Einstein, "the original theory of relativity is child's play."[7]

Einstein offered three tests to confirm or reject his theory. The first was the peculiar motion of Mercury. The entire orbit of Mercury was turning (precessing) more than Newton's law of universal gravitation could explain. It was a tiny but measurable amount: 43 seconds of arc a *century*. This unexplained advance in Mercury's *perihelion* (its closest point to the Sun) had given rise to a suspicion that one or more planets lay between Mercury and the Sun. The suspicion was so strong that this suspected planet, scorched by the nearby Sun, had already been given a name: Vulcan, after the Roman god of fire. Many observers had tried to spot it as a tiny dot passing across the face of the Sun, and some even claimed success. But when an orbit for the planet was calculated and its next passage across the face of the Sun was due, the planet never kept the appointment. It did not exist. But it was not until Einstein formulated the general theory of relativity that the 43-seconds-per-century anomaly could be explained.

Einstein was more than just pleased when he realized that his theory could account for this discrepancy in the motion of Mercury. "I was beside myself with ecstasy," he wrote.[8] This explanation of a perplexing problem gave general relativity high credibility. But the power of the general theory would be even more evident if it could predict something never before contemplated or detected.

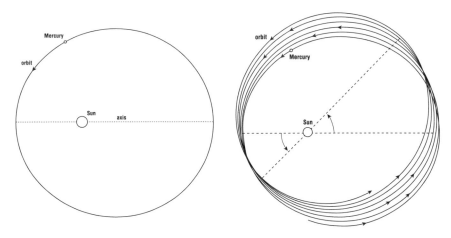

Precession of Mercury's orbital axis. *Left*: According to Newtonian theory, a planet orbiting the Sun follows a fixed elliptical path. *Right*: Einstein's general theory of relativity predicts that the axis of the ellipse will gradually rotate.

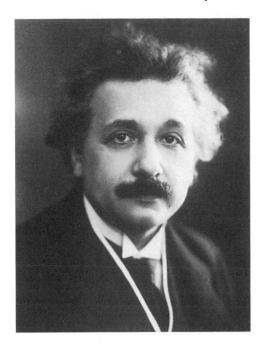

Albert Einstein in 1922. [Burndy Library & AIP Emilio Segrè Visual Archives]

Einstein offered two such predictions: that starlight passing close to the Sun would be bent, and that light leaving a massive object would have its wavelengths extended so that the light would be redder. This *gravitational redshift* was so small an effect that it could not be detected in the Sun with the equipment available at the time, so this proof of general relativity had to wait many years. It was finally detected in 1959 by Robert V. Pound and Glen A. Rebka, Jr., using the recently discovered Mössbauer effect in which the gamma rays emitted by atomic nuclei serve as the most precise of clocks. One of these atomic clocks was placed in the basement of a building and moved up and down so that its depth in the Earth's gravitational field varied minutely. The deeper in the basement the clock was, the longer the wavelength of its radiation. It had taken 45 years, but the gravitational redshift predicted by Einstein had at last been confirmed.[9]

In contrast, the gravitational deflection of starlight predicted by Einstein could be tested at most total eclipses of the Sun. Between 1911 and 1915, Einstein revised his calculation, using his new general theory of relativity. He found the deflection to be twice the initially assigned value. Starlight passing near the Sun would be bent 1.75 arc seconds. (Einstein and the world were fortunate that his initial prediction was not tested before it was revised; otherwise his later figure, although rigorously honest, might have seemed to be a manipulation to make the numbers come out right. There would have been far less drama in the confirmation of relativity.)

In 1916 Einstein published his complete general theory of relativity. But the world hardly noticed. For two years, nations had been locked in the First World War. Feelings against Germans ran strong in France and Great Britain, just as the Germans hated the French and British. Einstein sent his paper to a friend, scientist Willem de Sitter in the Netherlands. De Sitter, in turn, forwarded the paper to Arthur Eddington in England. At the age of 34, Eddington was already famous for his pioneering work in how a star emits energy. Eddington instantly recognized the significance of Einstein's discoveries and was deeply impressed by its intellectual beauty. He shared the paper with other scientists in Britain.

The 1919 Test

Frank Dyson, the astronomer royal of England, began planning for a British solar eclipse expedition in 1919 to test relativity. It was the perfect eclipse for the purpose, because the Sun would be standing in front of the Hyades, a nearby star cluster, so there would be a number of stars around the eclipsed Sun bright enough for a telescope to see. But the timing could hardly have been worse. In 1917 Britain was in the midst of a terrible war whose issue was still very much in doubt. Yet somehow Dyson managed to persuade the government to fund the expedition, despite the fact that its purpose was to test and probably confirm the theory of a scientist who lived in Germany, the leader of the hostile powers.

Meanwhile, American astronomers had an earlier opportunity to verify or disprove relativity. In the final months of World War I, a total eclipse passed diagonally across the United States from the state of Washington to Florida. On June 8, 1918, a Lick Observatory team led by William Wallace Campbell and Heber D. Curtis, observing from Goldendale, Washington, pointed their instruments skyward to render a verdict on relativity. The weather was mostly cloudy, but the Sun broke through for three minutes during totality. Measuring and interpreting the plates had to wait several months, however, until comparison pictures could be taken of the same region of the sky at night, without the Sun in the way. By this time, Curtis was in Washington, D.C. working on military technology for the government as the war came to a close. Curtis returned to Lick in May 1919 and the results were announced in June. The star images were not as pointlike as desired. Curtis could detect no deflection of starlight. By this time,

Arthur S. Eddington. [AIP Emilio Segrè Visual Archives, gift of S. Chandrasekhar]

word was spreading about the findings of the British expeditions, and the Lick paper was never published.[10]

Four months after the armistice, two British scientific teams were poised for departure. Andrew C. D. Crommelin and Charles R. Davidson, heading one party, were to set sail for Sobral, about 50 miles (80 kilometers) inland in northeastern Brazil. Eddington, Edwin T. Cottingham, and their team were headed for Principe, a Portuguese island about 120 miles (200 kilometers) off the west coast of Africa in the Gulf of Guinea. On the final day before sailing, the four team leaders met with Dyson for a final briefing. Eddington was extremely enthusiastic and confident that Einstein was right. A deflection of 1.75 arc seconds would confirm relativity. Half that amount—a deflection of starlight by 0.87 arc seconds—would reconfirm Newtonian physics.[11]

"What will it mean," asked Cottingham, "if we get double the Einstein deflection?"

"Then," said Dyson, "Eddington will go mad and you will have to come home alone!"

On Principe, worrisome weather conditions greeted Eddington and Cottingham. Every day was cloudy. However, May was the beginning of the dry season and no rain fell—until the morning of the eclipse. The fateful day, May 29, 1919, dawned overcast, and heavy rain poured down. The thunderstorm moved on about noon, but the cloud cover remained. The Sun finally peeped through 18 minutes before the eclipse became total but continued to play peekaboo with the clouds. Said Eddington: "I did not see the eclipse, being too busy changing plates, except for one glance to make sure it had begun and another halfway through to see how much cloud there was."

Eddington had much cause for worry. He was not interested in prominences or the corona. He needed to see clearly the region around the Sun. With great care, the plates were developed one at a time, only two a night. Eddington spent all day measuring the plates. Clouds had interfered with the view, but five stars were visible on two plates. When he had finished measuring the first usable plate, Eddington turned to his colleague and said, "Cottingham, you won't have to go home alone."

Eddington measured the displacement in the stars' positions, extrapolated to the limb of the Sun, to be between 1.44 to 1.94 arc seconds, for a mean value of 1.61 ± 0.30 arc seconds. The deflection agreed closely with Einstein's prediction. In later years, Eddington referred to this occasion as the greatest moment of his life.[12]

The expedition to Sobral had equally threatening weather, but there was a clear view of totality except for some thin fleeting clouds in the middle of the event. Crommelin and Davidson stayed in Brazil until July to take reference pictures of the star field without the presence of the Sun and then brought all their photographic plates back to Britain before measuring them. They found that their largest telescope had failed because heat caused a slight change in focus, which spoiled the pinpoint images of the stars. But the other instrument had worked well, and its plates also supported Einstein's prediction. Their mean value was 1.98 ± 0.12 arc seconds.

It was now September and no news about eclipse results had been released. Einstein was curious and inquired of friends. Hendrik Antoon Lorentz, the Dutch physicist, used his British contacts to gather news and telegraphed Einstein: "Eddington found star displacement at rim of Sun."[13] He also announced the favorable results at a scientific meeting in Amsterdam on October 25, with Einstein in attendance. But no reporters were present, and no word of the discovery was published.

Finally, on November 6, 1919, the Royal Society and the Royal Astronomical Society held a joint meeting to hear the results of the eclipse expeditions. The hall was crowded with observers, aware that an age was ending. From the back wall, a large portrait of Newton looked down on the proceedings. Joseph John Thomson, president of

Albert Einstein and Arthur S. Eddington in Eddington's garden, 1930. [Royal Greenwich Observatory]

the Royal Society and discoverer of the electron, chaired the meeting, praising Einstein's work as "one of the highest achievements of human thought." He called upon Frank Dyson to summarize the results and introduce the reports of the eclipse team leaders. The astronomer royal concluded his presentation by saying: "After careful study of the plates I am prepared to say that there can be no doubt that they confirm Einstein's prediction. A very definite result has been obtained that light is deflected in accordance with Einstein's law of gravitation."[14]

Einstein awoke in Berlin on the morning of November 7, 1919, to find himself world famous. Hordes of reporters and photographers converged on his house. He genuinely did not like this attention, but he found a way to turn the disturbance to the benefit of others. He told the reporters about the starving children in Vienna. If they wanted to take his picture, they first had to make a contribution to help those children. Suddenly Einstein was a celebrity.

More Tests

Expeditions from many nations set off to retest relativity by measuring the deflection of starlight during the eclipse of September 21, 1922. The shadow passed across Somalia, the Indian Ocean, and Australia. Most of the teams had bad luck; the weather failed them. But the successful American delegation, a Lick Observatory team

William Wallace Campbell about 1914. [Mary Lea Shane Archives of the Lick Observatory]

again headed by Campbell, observed the eclipse from Wallal on the northwest coast of Australia. After the Lick Observatory disappointments in 1914 and 1918, this expedition was an all-out effort, with the best of equipment and lots of rehearsal. The results had to wait, though, until comparison photographs had been taken and until Campbell and Robert C. Trumpler had measured the plates with utmost care. In April 1923, seven months after the eclipse, Campbell announced: "The agreement with Einstein's prediction from the theory of relativity . . . is as close as the most ardent proponent of that theory could hope for."[15] They measured the displacement at 1.72 ± 0.11 arc seconds.

When Campbell was asked for his personal reaction to this new confirmation of relativity, he replied: "I hoped it would not be true."[16] Campbell was not the only one reluctant to accept such a new and difficult concept. There were a number of scientists who did not share Eddington's delight in Einstein's victory.

For the next half century, most of the Sun's rendezvous with the Moon were attended by scientists carrying instruments to remeasure the deflection of starlight. A wide range of conflicting values resulted. For the eclipse of June 30, 1973, a definitive measurement was attempted. Once again, the eclipse was conducive for study: the second-longest totality of the twentieth century. A combined University of Texas and Princeton University team journeyed into the Sahara to watch the eclipse from an oasis in Mauritania where totality would last 6 minutes 18 seconds. They took special precautions about temperature variations, emulsion creepage, and other factors. The precautions were almost for naught. Eclipse day brought a cruel surprise. Before the Moon blocked the Sun's light, a dust storm provided its own eclipse, subsiding only a few minutes before totality, but leaving the atmosphere choked with dust that blocked out 82 percent of the light.[17] Still, the astronomers recorded 150 measurable images and found a deflection of 1.66 ± 0.18 arc seconds, well in accord with Einstein's prediction.

Even as that attempt was made, a new method of testing relativity by the deflection of radiation around the Sun was emerging. Radio astronomers were using widely separated radio telescopes to measure positions with an accuracy greater than optical telescopes. Quasars

Heber D. Curtis about 1913. [Mary Lea Shane Archives of the Lick Observatory]

The 1922 Lick Observatory eclipse expedition unloads equipment at Wallal, Australia. [Mary Lea Shane Archives of the Lick Observatory]

Makeshift mount for the principal telescope used by the Lick Observatory team to test relativity at the 1922 eclipse in Wallal, Australia. [Mary Lea Shane Archives of the Lick Observatory]

Women's work at the 1922 Lick Observatory eclipse expedition to Wallal, Australia. Note the mosquito netting. [Mary Lea Shane Archives of the Lick Observatory]

had been discovered and nearly all astronomers interpreted them to be the most distant objects in the universe—so distant that they were essentially fixed markers. Their angular distance from one another provided a new coordinate system against which the positions of all other objects could be referred. Because some of the quasars lay near the ecliptic, the Sun's apparent path through the heavens in the course of a year, the Sun's gravity would annually deflect the quasars' radiation, making them seem to shift slightly in position. Only this time, the radiation would be radio waves rather than visible light. And no eclipse was necessary because radio telescopes do not require darkness. In 1974 and 1975, Edward B. Fomalont and Richard A. Sramek used a 22-mile (35-kilometer) separation between radio telescopes to measure the deflection of light at the limb of the Sun as 1.761 ± 0.016 arc seconds, a result that not only confirmed relativity but also favored Einsteinian relativity over slightly variant relativity formulations by other scientists.[18]

Solar eclipse photographed by the Lick Observatory expedition at Wallal, Australia, September 21, 1922. [Mary Lea Shane Archives of the Lick Observatory]

Solar eclipses may no longer be the most accurate way of determining the relativistic deflection of starlight, but when the general theory of relativity needed proof from a phenomenon predicted but never before observed, it was solar eclipses that provided the first and most dramatic demonstration of Einstein's masterpiece and brought relativity and Einstein to the attention of the entire world.

9

Modern Scientific Uses for Eclipses

I have a little shadow that goes in and out with me,
And what can be the use of him is more than I can see.
—Robert Louis Stevenson (1885)

When scientists first turned their attention to solar eclipses, they put them to use to clock the motions of the Moon around the Earth and the Earth around the Sun. By trying to predict the exact time and location of the path of a solar eclipse, astronomers could take note of their errors and refine their knowledge of the orbits of the Earth and Moon. This work was pioneered by Edmond Halley in 1715 for a total eclipse crossing southern England.

More than a century later, in the early days of astrophysics, a second use for total eclipses emerged. The eclipse of 1842, carving a path across southern France and northern Italy, gave European scientists a front-row seat to see for themselves the occasionally reported effects surrounding totality: the corona, prominences, and chromosphere. They were awed and realized that the Sun, by covering its face, was revealing physical aspects of itself that were not visible at any other time. By studying the atmosphere of the Sun during total eclipses and the visible surface of the Sun at all other times, scientists began to perceive how hot the interior of the Sun must be.

Then, unexpectedly, early in the twentieth century, there arose a third great scientific use for solar eclipses: to confirm or deny the peculiar new theory of gravity and the structure of the cosmos offered by Albert Einstein. The first affirmative answer came from an eclipse in 1919, with data from subsequent eclipses over the remainder of the century adding to the certainty.

Meanwhile, the original scientific uses for total eclipses waned. The U.S. Naval Observatory no longer refines the orbits of the Moon or Earth from eclipse timings. Equipment and techniques developed since 1868 allow astronomers to study the prominences of the Sun independent of eclipses. In 1930, Bernard Lyot of France invented the coronagraph, a telescope with a special optical system to create an artificial eclipse so that the brighter regions of the corona can be studied without waiting for a precious few moments of eclipse totality in some remote corner of the world. And the relativistic bending of starlight can now be checked more precisely by using radio waves, which can be received during broad daylight, without waiting for an eclipse.

Have eclipses been used to exhaustion by scientists and then abandoned to the care of aesthetes and amateur astronomers?

Dogging the Sun's Diameter

David Dunham is not about to consign eclipses to the historical archives of scientific relics. He and his colleagues Joan Dunham, Alan Fiala, Paul Maley, amateur astronomers David Herald of Australia and Hans Bode of Germany, and other members of the International Occultation Timing Association (IOTA), journey to eclipses around the world for the express purpose of almost missing them—that is, missing what almost everyone else is most anxious to see: the longest possible duration of totality. You will never find Dunham and his colleagues along the center line. Instead, they and any friends and local people who wish to join them position themselves across the northern and southern limits of the eclipse path so that they witness just a few moments of totality, or none at all.

From their positions at the edges of totality, the top or bottom of the Moon just briefly covers the full face of the Sun. The corona appears and then fades away as the Sun reemerges. Here Dunham finds precisely the information he seeks. He and his coworkers are trying to measure minute changes in the diameter of the Sun, and have enlisted the Moon for their service. The size and distance of the Moon are well known. The distance of the Sun is known to great accuracy. Therefore, by measuring the size of the Moon's shadow, they can derive the size of the body responsible for that shadow: the Sun.

By stretching a team of observers perpendicular to the expected edge of the eclipse path, typically from 0.5 mile (0.8 kilometer) outside to 1.5 miles (2.4 kilometers) inside, they can determine where the actual edge of totality passes to within a hundred meters. This translates into the ability to measure the angular radius of the Sun to an accuracy of 0.04 arc second, or about 20 miles (30 kilometers). The Sun in 1983 was, according to their measurements, about 0.4 to 0.5 arc second larger than it was in 1979, which means that the Sun had increased in radius by about 180 miles (290 kilometers). However, the Sun seemed to be 0.5 arc second smaller in 1979 than it was in 1715 or even 1925, a decrease in the solar radius by about 230 miles (375 kilometers).[1]

It is a little surprising to think that our Sun might be expanding, shrinking, or pulsating. But then, it was a shock to people almost four centuries ago when Galileo used his telescope on the Sun, a celestial body thought to be "pure" and immune to change, and found spots on the Sun that appeared, changed, disappeared, and showed that the Sun was rotating. Over the centuries, scientists recognized more and more changes on the Sun: the shape of the corona, prominences, flares, the sunspot cycle. What is most surprising is not that the Sun changes but that changes on the Sun are not reflected more dramatically (and catastrophically) in short-term climate changes on Earth.[2]

Do the Dunhams, Fiala, and their colleagues regret missing the "main attraction" of a total eclipse? No, says Dunham, he actually prefers the view from the edge and chides "center line prejudice." He sacrifices a long look at the corona and the eerie twilight at the center of the eclipse path, but he gains a maximum-duration view of Baily's Beads, the chromosphere, and the shadow bands, which he sees without special effort at virtually every eclipse. Dunham has observed thirteen eclipses in this fashion and, funds permitting, he plans to attend many more.[3]

If this variation in the diameter of the Sun is real, and he and his colleagues think it is, they suspect that it may be a cyclic pulsation tied to the sunspot cycle of approximately 11 years. Why, Dunham does not know. He refers theoretical questions to others, such as Sabatino Sofia. Dunham concentrates on refining his equipment and techniques to hold experimental error to low enough levels so that his data will have clear meaning.

The task is simple in theory but very complicated in practice. The raw data—what was seen from each precisely marked viewing position at precisely what time—are just the start of the project. To compute the angular size of the Sun to see if it has changed, one must take into consideration a host of factors, the most complex of which is the landscape of the Moon. Because of the mountains and valleys on the Moon, its limb is jagged, producing Baily's Beads as totality nears and the only light from the Sun's photosphere that reaches the Earth is passing through the deepest valleys. But because the Moon has an elliptical and inclined orbit, the features visible along the Moon's edge change from one eclipse to the next.[4]

For the radius of the Sun to be measured accurately using the Moon's shadow during an eclipse, the exact height of the north and south limb features of the Moon as they appear during each specific eclipse must be calculated to high accuracy; otherwise the fraction-of-a-mile uncertainty in the bead-creating diameter of the Moon washes out the accuracy in the measurement of the Sun's radius. Crit-

---

The Corona, Eclipses, and Modern Solar Research

by Jay M. Pasachoff

The disk of the Sun is visible from the ground every clear day, but the solar corona is most clearly visible from the ground only during total eclipses. Scientists are interested in the corona for four major reasons.

First, we learn about how the Sun works. All the energy that leaves the Sun passes through the corona. Why the temperature of the corona is as high as 2 million degrees Celsius is still a major question, with theories of coronal heating via magnetic fields currently dominating.

Second, we study the solar corona to learn more about the Earth. The corona continually expands into space in the form of the solar wind, which envelops our planet. Thus, the Sun causes many effects on Earth: the aurorae, shortwave and CB communication blackouts, and surges on power lines that lead to outages. Changes in solar radiation may affect the Earth's climate on a timescale short enough to detect.

Third, the Sun is a rather typical star, so by studying the Sun we gain detailed close-up information that we can apply to the distant stars that are mere pinpoints of light in our largest telescopes. The Sun is therefore a key to understanding the universe.

Fourth, the Sun is a celestial laboratory where we learn basic laws of physics. The conditions on the Sun are often not duplicable on Earth. For example, the density of the solar corona is so low that it would be considered a fantastic vacuum in a laboratory on Earth. The corona is a hot plasma that the solar magnetic field directs into the beautiful streamers seen during eclipses. Thus the corona reveals the behavior of hot gas held in a magnetic field. This knowledge is useful to magnetic fusion research, which may one day provide energy on Earth. In the laboratory, we have trouble holding plasma together long enough for protons and deuterons to come sufficiently close together to fuse, releasing energy according to Einstein's

ics of solar radius variation claims say that practitioners are overestimating the accuracy of their measurements.[5]

If the Sun actually does pulsate slightly in the course of a sunspot, and hence magnetic, cycle, it will probably take another decade or two to prove the point—a long-term project that brings to mind the cartoon of the turtle family, with the young turtle a few steps distant from his parents. "Don't look now," says Mother Turtle, "but I think Junior is running away again. The next few weeks will tell."[6]

Across the Spectrum

Dunham and IOTA members are by no means the only astronomers who still see important scientific value in solar eclipses. In fact, solar

---

formula $E = mc^2$. Positively charged particles repel each other strongly. In a hot gas, particle velocities are so high that protons and deuterons approach each other closely despite their mutual repulsion. But no material container can hold such hot gas, so magnetic fields are used to constrain it. Astronomical studies of the corona help physicists and engineers with this important problem.

The corona is so important to science that no method of observing it should be ignored. Solar telescopes in space, such as the Yohkoh satellite, the Solar and Heliospheric Observatory (SOHO), and the Transition Region and Coronal Explorer (TRACE) observe the Sun in parts of the spectrum that do not reach the ground and have provided magnificent new insights. But to limit the scattering of sunlight, space-borne coronagraphs to date have had to block out the innermost corona. And ground-based observations of the corona outside of eclipses cannot see the corona as far from the Sun as can be seen at a total eclipse. Thus, eclipse observations are essential supplements to ground- and space-based coronal observations if the full picture is to be grasped.

The traditional advantages of eclipses over space observations for solar research are flexibility and cost. New observations and instruments can be incorporated into an eclipse expedition on short notice. State-of-the-art and bulky equipment can be transported to eclipse sites far more cheaply than to space. Eclipse equipment does not have to meet launch standards of sturdiness. Further, eclipse instruments can be mounted on steady bases, with the Earth as a platform, and can be adjusted at the last minute by qualified scientists.

The SOHO solar telescope cost hundreds of millions of dollars. For less than one-tenth of one percent of these costs, a well-equipped modern ground-based eclipse expedition can be mounted. Even allowing for some of the eclipses to be clouded out, eclipses are a very cost-efficient way of doing astronomy research.

eclipses provide so many unique opportunities to study the Sun's atmosphere that this chapter can present only a small sampling of the research in progress.

Spectroheliographs and coronagraphs allow photographic inspection of many aspects of the Sun's chromosphere, prominences, and lower corona outside of an eclipse. But observing in special wavelengths or by placing an occulting disk at the focal plane to block the face of the Sun cannot compensate for the turbulence in the Earth's atmosphere that causes the sunlight to be refracted so that it comes in from very slightly different angles. This deflection of the light blurs the image enough to suppress viewing of fine detail in the corona and prominences altogether, leaving only the most conspicuous solar atmospheric features visible.

For more than a century, astronomers have used occultation by the sharp lunar limb to detect detail in sources that is much too fine to resolve with the most powerful telescopes. What appears to the telescope as a single star can be identified as a close binary star system as the Moon passes in front of it.

The basis of this technique is to measure the "occultation curve" of the source— how rapidly the intensity of the source falls off as the solid surface of the Moon occults it. If the source is relatively broad and diffuse, the intensity decreases relatively gradually, over many seconds, as the Moon slides in front of it and covers it up. If the source is extremely sharp and fine, the intensity drops to zero in a few seconds or less. Eclipses allow scientists to apply the occultation technique directly to the solar limb, resolving the density and temperature profile of the solar chromosphere in fine detail.

Alan Clark and John Beckman pioneered this technique in the infrared wavelengths, observing solar eclipses with telescopes aboard high-flying aircraft. Charles Lindsey and Eric Becklin refined their techniques, using NASA's Kuiper Airborne Observatory, a military cargo jet converted to carry a 36-inch (0.9-meter) infrared telescope.

Lindsey, Becklin, and their colleagues from the United States, the United Kingdom, and Canada applied the same technique to the spectacular eclipse of July 11, 1991, which passed directly over Hawaii's Mauna Kea Observatory, perched at an elevation of 13,860 feet (4,225 meters). Because of exceptionally dry conditions atop high mountains, infrared wavelengths can be recorded there without the use of aircraft. These observations were made with the 50-foot-diameter (15-meter-diameter) James Clerk Maxwell Telescope, a radio observatory operated on Mauna Kea by the United Kingdom, the Netherlands, and Canada.

Occultation curve observations made during solar eclipses in the 1980s and early 1990s have now provided scientists with remarkably detailed profiles of the structure of the solar chromosphere in radiation ranging from 30 out to 1,200 microns (1.2 millimeters), wavelengths that are about 40 to 1,600 times longer than the reddest light the human eye can see.[7]

The Baked Alaska Problem

Spectroscopy of the lower atmosphere of the Sun reveals the presence of carbon monoxide (CO)—a molecule. That is weird. Temperatures near the surface of the Sun should instantly split carbon monoxide into carbon and oxygen unless this molecular gas has a temperature as low as sunspots, some 2,700°F (1,500°C) cooler than the 10,000°F (5,500°C) photosphere.

Exactly where is this "cold" material in the solar atmosphere and how high does it extend? Attempts to pin down its precise vertical extent had been thwarted by fluctuations in the solar image caused by the Earth's turbulent atmosphere. This turbulence causes stars to appear to twinkle and the Sun's disk to appear to quiver.

Alan Clark thought it might be possible to avoid this "seeing" problem and pinpoint the position of the carbon monoxide by using the geometry of a solar eclipse. He and Rita Boreiko made initial observations at eclipses in the 1980s by flying in a small Learjet above most of the infrared-radiation-absorbing water vapor in the Earth's atmosphere. They got mixed results.

But when 1991 brought a total eclipse across the summit of Mauna Kea in Hawaii, Clark and David Naylor exploited the opportunity by using infrared photometry to show that the carbon monoxide lay in a narrow band above the photosphere.

The annular eclipse of May 10, 1994, however, provided Clark with an almost ideal opportunity to pinpoint the height of this surprisingly cold atmospheric component. The Moon's shadow swept close to the world's largest solar telescope, the McMath-Pierce instrument on Kitt Peak in southern Arizona. Furthermore, this telescope was equipped with a powerful infrared spectrograph and imaging camera. As the Moon progressively eclipsed the solar atmosphere, Clark, Charles Lindsey, Douglas Rabin, and William Livingston were able to watch specific spectral lines of carbon monoxide molecules as they changed from absorption lines when seen against the hotter background of the solar disk to emission lines when seen against the background of space at the solar limb. They could determine the position of this process to within about 50 miles (80 kilometers). Such observations are relatively unhindered by "seeing" fluctuations because the obscuring Moon is above the Earth's atmosphere. Most of the carbon monoxide was concentrated within 280 miles (450 kilometers) of the Sun's surface, although very small amounts were detected up to about 625 miles (1,000 kilometers).

The remaining problem is to explain how this cool layer can exist, in which carbon monoxide survives dissociation into atoms while surrounded by fierce temperatures. Clark calls this the "baked Alaska problem." How can the ice cream (the carbon monoxide) remain frozen within the cake while in an oven (the chromosphere) hot enough to bake the meringue?

The present idea is that a network structure of hot, concentrated plasma containing magnetic fields is spread like a fishnet across the solar surface. The carbon monoxide appears to survive in cold pools within the cells of this network, cooling itself by emitting intense infrared radiation by which it is detected along the solar limb. It survives only up to the altitude at which the network magnetic fields spread out to cover the solar surface uniformly. These measurements using eclipses thereby establish the "canopy height" for the chromospheric network magnetic fields.

Thus, eclipses provide the resolution necessary to help clarify the contribution of magnetic fields to the structure of the Sun's lower atmosphere.[8]

More Information from the Infrared

Jeffrey Kuhn and his colleagues also use eclipses to make observations of the solar atmosphere in infrared wavelengths. For the 1998 eclipse, they instrumented a C-130 cargo plane with a hole in its roof through which they could track the Sun. This National Center for Atmospheric Research aircraft allowed them to climb above much of the water in the atmosphere that absorbs the Sun's infrared radiation. Kuhn's team flew from Panama out over the Pacific Ocean to intercept the eclipse. By flying eastward along the eclipse path, they were able to extend the duration of totality from 4 minutes to 5. Through these observations, Kuhn and his coworkers discovered a new line in the spectrum of the Sun created by silicon with 8 of its 14 electrons missing. This spectral line is strong enough to use in future eclipses as a probe to trace and measure the magnetic fields of the corona.[9]

Infrared studies of the Sun's corona during eclipses have also allowed Kuhn and his colleagues to detect fine dust particles deposited in the inner solar system by outgassing comets and colliding asteroids. Sunlight reflecting from this dust provides the faint, hazy zodiacal light that can be seen under dark-sky conditions rising before the Sun or setting after the Sun. The dust that reflects the zodiacal light lies primarily in and near the plane of the system and gradually spirals in closer to the Sun. With their infrared measurements during eclipses, Kuhn and his coworkers have detected zodiacal dust within the corona, but, as it spirals into the Sun, this dust does not pile up in rings in the outer corona, as some astronomers had suspected.[10]

Threads and Plasmoids

Serge Koutchmy and Laurence November wait for the great natural coronagraph-in-space, the Moon, to eclipse the Sun to permit them a view of the lowest regions of the corona. They hunt for narrow, bright coronal loops indicative of powerful magnetic fields. Studying these loops carefully in ordinary white light, they have observed bright and dark "threads" within the loops—and have found that the dark threads must be essentially vacuums, void of material. Full evacuation requires a magnetic field of a specific strength. The evacuated dark threads provide November with a new perspective on how the magnetic field may determine the temperature of the corona.[11]

When the total solar eclipse of July 11, 1991 passed over Hawaii and the huge observatory complex atop the extinct volcano Mauna Kea, Koutchmy, November, and their coworkers seized the opportunity to observe the corona with the largest optical telescope ever to record a total eclipse, the 142-inch (3.6-meter) Canada-France-Hawaii Telescope. With so large a telescope, they were able to see details in the corona too small to be seen before. They think they observed a coronal plasmoid, a high-density bubble of ionized gas in the corona, about 900 miles (1,400 kilometers) in diameter, moving at 60 miles per second (100 kilometers per second) outward from a magnetic loop prominence that probably spawned it. During the four minutes of the eclipse, the plasmoid distorted, stretched out, and broke into smaller plasmoids. Koutchmy, November, and their colleagues suspect that it is these plasmoids, small-scale ejection events, not the large coronal mass ejections, that supply most of the material to the upper corona that the Sun loses to space as the background solar wind.[12]

Rockets and Eclipses

During solar eclipses long ago, the Chinese would beat drums, clang cymbals, and fire off skyrockets to scare away the dragon or dog that was eating the Sun. In modern times, most recently in 1988 and 1994, Gary Rottman, Frank Orrall, and Don Hassler have fired off rockets during solar eclipses, but their intent was discovery.

They used NASA Black Brant suborbital rockets to carry their ultraviolet cameras and spectrometers to an altitude of about 200 miles (325 kilometers), where extremely short ultraviolet wavelengths from the Sun's corona can be recorded before they are absorbed in the Earth's atmosphere. Because Rottman, Orrall, and Hassler were collecting the shortest ultraviolet wavelengths, they could have launched rockets to make these observations of the corona without waiting for a total eclipse. The high temperature of the corona gives it a brightness at very short wavelengths that overwhelms the brightness of the comparatively cool photosphere, allowing the corona to be seen even though the disk of the Sun is in full view. But total eclipses are the best time for rocket flights because the state of the corona they record in their five minutes above the Earth's atmosphere can be compared with the corona seen nearly simultaneously from the ground in visible and radio wavelengths, greatly increasing the scientific value of all the observations. Each wavelength provides its own revelations about the corona and, studied together, they provide powerful tests of the theories that attempt to explain peculiar features of the corona: its chemical abundances, its density irregularities (clumpiness), how it is heated to exceptionally high temperatures, and the mechanism by which the Sun drives the solar wind.

The Chaos of the Corona

The atmosphere of Earth is controlled by three primary forces: gravity, pressure variations, and the Earth's rotation. The Sun's atmosphere is shaped not only by gravity, pressure, and rotation, but also by a fourth force: magnetism. A total eclipse offers the best chance to see how these magnetic fields sculpt the atmosphere of the Sun. By revealing the corona, totality also provides a chance to study a high temperature plasma under conditions that cannot be duplicated on Earth: extremely hot ionized gases at extremely low density.

A total eclipse offers as well an opportunity to see how the corona changes with time and to what degree these fluctuations correlate with magnetic activity. Richard R. Fisher is one of a team of scientists who like to use eclipses to record the temperature and structure of

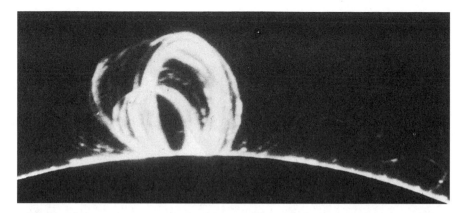

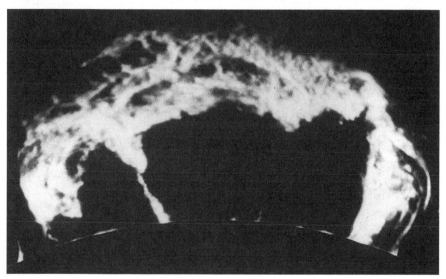

(top) Loop prominence. [National Solar Observatory/Sacramento Peak, NOAO]

(bottom) Eruptive prominence. [National Solar Observatory/Sacramento Peak, NOAO]

the corona with far more precision than observations of the Sun outside of eclipses provide.

Another astronomer still using eclipses to probe the Sun is Jay Pasachoff. He is eager to understand how the corona is heated to temperatures more than 300 times hotter than the photosphere. The energy of the Sun is created at its core, where the temperature is about 27 million °F (15 million °C). By radiation and then convection the photons of light energy make their way to the surface, gradually cooling en route, so that the Sun's photosphere glows with a temperature of only about 10,000°F (5,500°C). The base of the chromosphere, the lower atmosphere of the Sun, is cooler still, about 7,200°F (4,000°C). But the temperature in the corona exceeds one million degrees. Temperature is a measure of the random motion of atoms and subatomic particles, and the particles in the corona are indeed traveling at great speed. But the corona has a very low density, so the total amount of heat in the corona is quite small.

What causes the temperature in the corona to be so much higher than that of the visible surface of the Sun? These surprising temperatures seem to be caused by magnetic fields. The magnetic fields that cause sunspots in the photosphere pervade the corona as well. They are responsible for streamers and other coronal features; they also shape the prominences and filaments.

A favored explanation for the high temperature of the corona is that magnetic waves (for example, Alfvén waves), excited by motion in the convective region beneath the photosphere, propagate energy into the corona, where it is then dissipated into heat. Such waves might manifest themselves as local oscillations in coronal brightness with periods in the range of about $\frac{1}{10}$ to 10 seconds.

Jay Pasachoff and his students search the corona during total eclipses, such as the one of February 26, 1998, on Aruba. Some coronal heating theories predict the existence of waves along coronal loops. If magnetic waves are present, observers of the loops should see rapid oscillations in the density of the plasma (ionized gases) as the waves pass through the gases. Because the coronal gases are so hot, they give off light at distinctive wavelengths (such as a green spectral line that iron emits when half its 26 electrons are missing).

The coronal plasma is quite rarefied, however, so the glow is weak. But if waves pass through a point within these gases, the waves briefly increase the gas density at that point, thereby briefly increasing the brightness of the gas, and this fluctuation in brightness would give evidence that, indeed, waves are present.

Pasachoff and his students use the total phase of a solar eclipse to make a sequence of images of coronal loops using the light of the coronal green line. They watch for periodic brightness variations in small regions of the loops that would suggest the passage of waves. They think they may have detected oscillations with periods in the 0.2–5-second range. The analysis of data continues.

Pasachoff uses every accessible total eclipse to search for these oscillations in the corona of his favorite star. He hopes this research will help to identify what kind of wave is responsible for the surprisingly high temperatures of the corona and how that heating occurs.[13]

10

Observing a Total Eclipse

Now eclipses are elusive and provoking things . . . visiting the same locality only once in centuries. Consequently, it will not do to sit down quietly at home and wait for one to come, but a person must be up and doing and on the chase.

—Rebecca R. Joslin (1929)

A total eclipse of the Sun. What is this sight that lures people to travel great distances for a brief view at best and a substantial possibility of no view at all, with no rain check? And how do you get the most out of the experience of a total eclipse?

In these pages, eighteen astronomers, professionals and amateurs, plus some eclipse seekers from earlier times share their experiences with you. Among them, they have witnessed two hundred total eclipses.

"It's like a religious experience," says Jay Anderson, "the anticipation as the time until totality is counted in days, then hours, then minutes. It's the perfect buildup. Spielberg couldn't do it better. It's an intensely moving event."

Steve Edberg agrees: "It is the intensity of the event. You grab as much as you can. I like action in the heavens, and you can't get much better than this."

Roger Tuthill has a ritual he follows the day before an eclipse. He walks around inside the path of totality observing the people who live there. They have no idea what they are about to experience, and there are no words to adequately prepare them.

Panel of Eclipse Veterans

Jay Anderson, meteorologist, Environment Canada, Winnipeg

John R. Beattie, typesetter, New York City

Richard Berry, former editor-in-chief, *Astronomy*

Dennis di Cicco, associate editor, *Sky & Telescope*, Cambridge, Massachusetts

Stephen J. Edberg, astronomer, Jet Propulsion Laboratory, Pasadena, California; and past president, Western Amateur Astronomers

Alan D. Fiala, chief, Nautical Almanac Division, U.S. Naval Observatory, Washington, D.C.

Ruth S. Freitag, senior science specialist, Library of Congress, Washington, D.C.

Joseph V. Hollweg, professor of astronomy, University of New Hampshire, Durham

George Lovi, astronomy author/columnist, Lakewood, New Jersey (George died in 1993)

Frank Orrall, professor emeritus of physics and astronomy, Institute for Astronomy, University of Hawaii, Honolulu

Jay M. Pasachoff, Field Memorial professor of astronomy and director of the Hopkins Observatory, Williams College, Williamstown, Massachusetts

Leif J. Robinson, editor in chief, *Sky & Telescope*, Cambridge, Massachusetts

Virginia and Walter Roth, proprietors, Scientific Expeditions, Inc., Venice, Florida

Roger W. Tuthill, president, Roger W. Tuthill, Inc., telescope accessories, Mountainside, New Jersey

Jack B. Zirker, astronomer, National Solar Observatory, Sunspot, New Mexico, now retired

First Contact

There is a special feeling from the instant when the Moon begins to slide in front of the Sun. In less than a minute, observers with small telescopes see the first tiny "bite" out of the western side of the Sun. For Virginia Roth, "that's when the magic starts."

First contact remains a very special moment for Jay Pasachoff. As an astronomer, he can more readily than most appreciate all the factors that go into predicting precisely when an eclipse will occur and exactly where on Earth it will be seen. And when the call "First

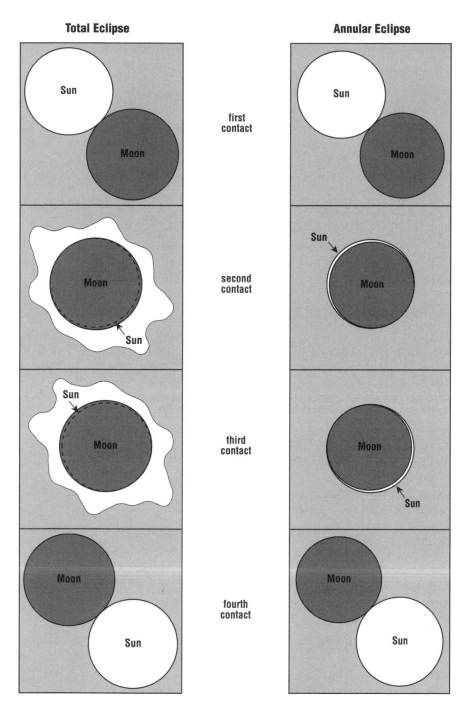

Points of contact in a total solar eclipse *(left)* and an annular eclipse *(right)*. In a partial eclipse, there are only first and fourth (last) contacts.

Contact" comes right on schedule, he always finds this ingenuity of mankind astounding.

Even a century and a half ago, this commencement of a rare event was already exerting a powerful effect on its beholders. French astronomer François Arago observed the 1842 eclipse from Perpignan in southern France amid townspeople and farmers who had been educated about the eclipse and who were watching the sky intently. "We had scarcely, though provided with powerful telescopes, begun to perceive a slight indentation in the sun's western limb, when an immense shout, the commingling of twenty thousand different voices, proved that we had only anticipated by a few seconds the naked eye observation of twenty thousand astronomers equipped for the occasion, and exulting in this their first trial."[1]

The Crescent Sun

Solar physicist Joe Hollweg likes to view the partial phases leading up to a total eclipse with a patch over one eye so that it becomes dark-adapted.[2] Totality is about as bright as a full-moonlit sky, and a dark-adapted eye makes for better viewing of totality.

The partial phase of a total eclipse has a power all its own, as the Moon steadily encroaches upon the Sun, covering more and more of its face. A partial eclipse close to total visited a Russian monastery in medieval times, near sunset on May 1, 1185. A chronicler recorded: "The sun became like a crescent of the moon, from the horns of which a glow similar to that of red-hot charcoals was emanating. It was terrifying to men to see this sign of the Lord."[3]

As the partial phase proceeds, a thousand tiny images of the crescent Sun may be visible on the ground beneath trees as the spaces between the leaves create pinhole cameras to focus the crescent image of the waning Sun.

Yet, as George B. Airy, astronomer royal of England, learned when he saw his first total eclipse in 1842: "[N]o degree of partial eclipse up to the last moment of the sun's appearance gave the least idea of a total eclipse."[4]

The Changing Environment

Every eclipse veteran urges newcomers to pause as totality approaches to look around at the landscape and notice the changes in light levels and color. Many people are surprised how little the landscape darkens until the last ten minutes or so before the eclipse becomes total. All the better for the drama, because once the light begins to fade, it sinks quite noticeably. Usually, everyone

Crescent images of the partially eclipsed Sun formed by foliage. [Mabel Loomis Todd: *Total Eclipses of the Sun*]

becomes quite silent. You can feel the tension and rising emotion. A primitive portion of your brain tugs at you to say that something peculiar is going on, that it ought not to be growing dark in the midst of the day.

Dennis di Cicco remembers that during the 1976 eclipse in Australia, the birds responded to the fading light by raising a racket and going to roost. In 1973, he witnessed an annular eclipse in Costa Rica that took place shortly after sunrise. The cows grazing around him ignored the partial phases of the eclipse, but as the eclipse reached maximum, the cows ceased grazing, formed a line, and marched back to their barn.

Walter Roth recalls watching the 1973 eclipse from a game preserve in Africa. As the light faded in the minutes just before totality, the birds flocked into the trees, complaining madly. Elephants, which had been grazing peacefully, milled around nervous and confused.

"As the light fades, the Sun is a thin crescent and shrinking quickly," says Richard Berry. "The quantity and quality of the light have begun to change noticeably. The temperature is dropping; the air feels still and strange."

Shadow Bands

As the eclipse nears totality and shortly after it emerges from totality, shadow bands—faint undulations of light rippling across the ground at jogging speed—are sometimes visible. They are one of the most peculiar and least expected phenomena in a total eclipse. Many eclipse veterans have never seen them; some do not want to take time to try because they occur in the last moments before totality as Baily's Beads and other beautiful sights are visible overhead. Other veterans consider shadow bands one of the true highlights of a total eclipse. They resemble the graceful patterns of light that flicker or

---

Catching Shadow Bands

by Laurence A. Marschall

Even though shadow bands are only visible for a few fleeting minutes, it is possible to catch them in flight if you prepare in advance. Get a large piece of white cardboard or white-painted plywood to act as a screen—the bands are subtle and can be more easily seen against a clean, white surface. A large white sheet staked to the ground may be more portable and will serve in a pinch, but ripples in the sheet can mask the faint gradations of the shadow bands.

Lay out on the screen one or two sticks marked with half-foot intervals (yardsticks will do nicely). Orient the sticks at right angles to one another so that at the first sign of activity you can move one stick to point in the direction that the shadow bands are moving. Then, using the marks on that stick, make a quick estimate of the spacing between the bands (typically 4 to 8 inches, or 10 to 20 centimeters). Finally, using a watch, make a quick timing of how long a bright band takes to go a foot or a yard. Jot down the figures or, better yet, dictate your measurements into a small tape recorder. If you practice this procedure before the eclipse, you will be able to see the shadow bands and then swing your attention back to the sky to catch Baily's Beads and the onset of totality.

The second stick, by the way, is reserved for measuring shadow bands *after* totality, if they are visible.

Later, when the excitement of the eclipse is over, you can take stock of the data. Do the shadow bands seem to move at all? At some eclipses, especially when the air is very still, they just shimmer without going anywhere. If they move, how fast? Typical speeds are about 5 to 10 miles per hour (8 to 16 kilometers per hour). Do they change directions after eclipse? Usually they do, unless you happen to be standing directly along the centerline of totality.

Shadow bands on an Italian house in 1870. [Mabel Loomis Todd: *Total Eclipses of the Sun*]

glide across the bottom of a swimming pool. Shadow bands occur when the crescent of the remaining Sun becomes very narrow so that only a thin shaft of sunlight enters the atmosphere of Earth overhead. There it encounters currents of warmer and cooler air that have slightly different densities. The different densities act as very weak lenses to bend the light passing from one parcel to the next. It is the focusing and defocusing of light by these ever-present air currents as they are blown about by the wind that causes stars to twinkle. The Sun would twinkle too if it were a starlike dot in the sky. And so it does, near the total phase of a solar eclipse. When the Sun has narrowed from a disk to a sliver, the light from the sliver twinkles in the form of shadow bands rippling across the ground.

In 1842, George B. Airy, the English astronomer royal, saw his first total eclipse of the Sun and recalled shadow bands as one of the highlights: "As the totality approached, a strange fluctuation of light was seen . . . upon the walls and the ground, so striking that in some places children ran after it and tried to catch it with their hands."[5]

The Approach of Totality

During the final one to two minutes before totality, the Sun is about 99 percent covered and the light is fading rapidly. In these brief moments, Baily's Beads, the Diamond Ring, and the corona all appear.

Looming on the western horizon, growing ever larger, is the Sun-cast dark shadow of the Moon. And it is coming toward you. Alan

Diamond Ring Effect,
June 11, 1983. [© 1983
Stephen J. Edberg]

Fiala describes its appearance as the granddaddy of thunderstorms, but utterly calm. If you are observing from a hill with a view to the west, the approach of the Moon's shadow can be quite dramatic, even chilling. In the words of astronomer Mabel Loomis Todd a century ago, it is "a tangible darkness advancing almost like a wall, swift as imagination, silent as doom."[6] And this is how astronomer Isabel Martin Lewis in 1924 described the onrush of the Moon's shadow at the onset of totality: "[W]hen the shadow of the moon sweeps over us we are brought into direct contact with a tangible presence from space beyond and we feel the immensity of forces over which we have no control. The effect is awe-inspiring in the extreme."[7]

As the shadow races east toward you, it accelerates. In the last few seconds before totality, you feel as if you are being swallowed by some gigantic whale. In these last few seconds before totality, as the soft white glow of the corona begins to silhouette the dark disk where the face of the Sun once shone, the ends of the remaining sliver of the Sun fracture into Baily's Beads. Each bead lasts only an instant and flickers out as new ones form. As the length of the Sun's sliver shortens, the two separated groups of beads converge and combine, and suddenly only one remains—the Diamond Ring. "For one fleeting moment this last bead lingers, like a single jewel set into the arc that is the lunar limb," says John Beattie. And then it is gone.

The eclipse is total.

Corona Emerging/Second Contact

It is difficult to find words that do justice to the corona, the central and most surprising of all features of a total eclipse. The Sun has vanished, blocked by the dark body of the Moon. A black disk surrounded by a white glow. "It is the eye of God," says Jack Zirker.

In the bitter cold of the high plateau in Bolivia, astronomer Frank Orrall was part of a research team studying the total eclipse of 1966. In reviewing his observing notes, he realized that he had written "The heavens declare the glory of God." "My notebooks normally do not say such things," he mused.

Even the color of the corona is difficult to capture. "I prefer 'pearly,'" says Ruth Freitag, "because it conveys a luminous quality that 'white' lacks." "A silver gossamer glow," suggests Steve Edberg. "Remember," says George Lovi, "the human eye is the only instrument that can see the corona in all its splendor."

During Totality

In the midst of totality, it is particularly impressive to look around at the nearby landscape and the distant horizon. Depending on your position within the eclipse shadow, the size of the shadow, and cloud conditions away from the Sun's position, the light level during totality can vary greatly from eclipse to eclipse and from place to place within an eclipse. On some occasions, totality brings the equivalent of twilight soon after sunset; on others it is dark enough to make reading difficult. Yet it is never as black as night. The color that descends upon everything is hard to specify. Most veteran observers describe it as an eerie bluish gray or slate gray, and then apologize for the inadequacy of the description. The Sun's corona contributes some brightness, but only about as much as a full moon. Primarily, it is light reflected from a short distance away, where the Sun is not totally eclipsed, that brightens your location within the shadow of the Moon.

Looking around the horizon in the midst of totality, you have the best impression of standing in the shadow of the Moon. You also have a renewed appreciation for the power of the Sun. There it is, its face completely blocked by the Moon. How big in the sky is that body whose overwhelming brightness creates the day and banishes the stars from view? Reach out your arm toward the eclipsed Sun, just as you did to measure the Moon. Squeeze the darkened Sun between your index finger and your thumb until it just fits between them. It is the size of a pea. Yet that little spot in the sky, darkened now, is usually enough to blanket the Earth in light and warmth and completely dazzle the eye. For just a few minutes its face is hidden. Daytime has become night and the temperature is falling. Do you feel something of what ancient people must have felt as they

Palm trees reach toward
the eclipsed Sun in Papua New
Guinea, November 22, 1984.
(Nikon FE, 24 mm focal length,
f/2.8, 1/4 second, with ISO 200
film. © 1984 Jay M. Pasachoff]

watched the Sun, upon which they depended for warmth and light
and life itself, disappear?

Leif Robinson emphasizes that a total eclipse cannot be experienced vicariously. Even the best photography and videotapes are pale
reflections of the event. Taking pictures during an eclipse is fine, but
be sure you *look* at what is truly a visual spectacle. Do not miss the
ambience of the moment either, he advises. Look at other people to
see how they are reacting. Notice the changing colors in the sky. He
does not, however, recommend trying to see planets or stars in the
sky during a total eclipse. "Time during totality is too precious to
spend straining to see stars when you can see those same objects
much better in the nighttime sky." Steve Edberg feels differently:
"One of my strongest memories from eclipses is seeing stars during
the day."

The transformation in the appearance of the Sun from a bright
crescent to a dark disk surrounded by ghostly light was recorded by
François Arago, the dean of French astronomy a century and a half
ago, as he watched the eclipse of July 8, 1842, with a group of
astronomers and nearly 20,000 townspeople in southern France.

[W]hen the sun, being reduced to a narrow filament, began to throw
only a faint light on our horizon, a sort of uneasiness took possession

The View from the Edge

by Alan D. Fiala

The advice in this chapter assumes that the observer is near the central line of an eclipse. But observing a total eclipse from the center of its path is not the only possibility.

At any total eclipse, if you move toward either edge of the path of totality, you gain in the duration of Baily's Beads activity. The number of beads you see depends on whether you go toward the southern or northern limit of the eclipse path. The southern edge provides more beads because the terrain of the Moon's southern limb is much rougher. The bead activity at the edges of the path is always caused by the same lunar features, so you can control the amount of bead activity you see by how close you come to the correctly predicted limit.

If you observe from the centerline of the eclipse, the beads are confined to the eastern limb of the Moon as totality approaches and the western edge of the Moon after the end of totality. Bead activity at the eastern and western limbs is strongly affected by lunar librations,[8] which change the view significantly from one eclipse to another and determine whether or not at second or third contact there is any Diamond Ring Effect at all.

If you want to study the inner corona or the corona near one pole (where the corona tends to appear more brushlike), once again you gain by going toward the edge of the path of totality. In this way, you control what portion of the inner corona you observe.

The centerline is not the only place from which to enjoy a total eclipse of the Sun.

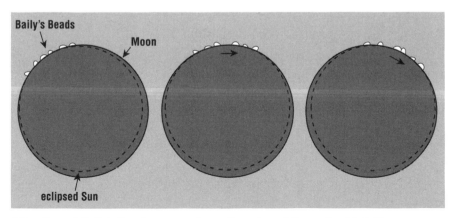

Migration of Baily's Beads as seen from the edge of the path of totality.

of every breast, each person felt an urgent desire to communicate his emotions to those around him. Then followed a hollow moan resembling that of the distant sea after a storm, which increased as the slender crescent diminished. At last, the crescent disappeared, darkness instantly followed, and this phase of the eclipse was marked by absolute silence. . . . The magnificence of the phenomenon had triumphed over the petulance of youth, over the levity affected by some of the spectators as indicative of mental superiority, over the noisy indifference usually professed by soldiers. A profound calm also reigned throughout the air: the birds had ceased to sing.

After a solemn expectation of two minutes, transports of joy, frenzied applauses, spontaneously and unanimously saluted the return of the solar rays. The sadness produced by feelings of an undefinable nature was now succeeded by a lively satisfaction, which no one attempted to moderate or conceal. For the majority of the public the phenomenon had come to a close. The remaining phases of the eclipse had no longer any attentive spectators beyond those devoted to the study of astronomy.[9]

How do you take it all in? It is not possible. There is too much to see—and feel. "Everybody, myself included, tries to do too much," says Steve Edberg. "Save time near the beginning, middle, and end of totality just to stare," urges Jay Anderson. "Make a deliberate effort to store the sights in your mind." Mabel Loomis Todd wrote, a century ago, that "when Dr. Peters of Hamilton College was asked what single instrument he would select for observing an eclipse, he replied, 'A pillow.'"[10]

The End of Totality

All too soon, no matter what the duration of totality (and it can never exceed 7 minutes 31 seconds), the Moon's shadow moves on and a bright flash of sunlight appears from the western edge of the Moon.

Rebecca Joslin and her college astronomy teacher traveled from the United States to Spain for an eclipse in 1905, only to be clouded out. So they shifted their attention to nature around them until light pierced the cloud to tell them that the unseen total phase of the eclipse was over.

But we hardly had time to draw a breath, when suddenly we were enveloped by a palpable presence, inky black, and clammy cold, that held us paralyzed and breathless in its grasp, then shook us loose, and leaped off over the city and above the bay, and with ever and ever increasing swiftness and incredible speed swept over the Mediterranean and disappeared in the eastern horizon.

Shivering from its icy embrace, and seized with a superstitious terror, we gasped, *"what* was *that?"*. . . The look of consternation on M's face lingered for an instant, and then suddenly changed to one of radiant joy as the triumphant reply rang out, *"that* was the *shadow* of the *moon!"*[11]

"I doubt if the effect of witnessing a total eclipse ever quite passes away," wrote Mabel Loomis Todd. "The impression is singularly vivid and quieting for days, and can never be wholly lost. A startling nearness to the gigantic forces of nature and their inconceivable operation seems to have been established. Personalities and towns and cities, and hates and jealousies, and even mundane hopes, grow very small and very far away."[12]

"Beware of post-eclipse depression," warns Steve Edberg. One minute after third contact, with the passage of the Moon's shadow, the disappearance of the shadow bands, and the reappearance of the crescent Sun, you feel exhausted: worn out by totality and the excitement of getting ready for it. Most observers are too tired to watch after third contact, and what follows is anticlimactic anyway. The cure for blue sky blues? "Socialize," says Edberg. "Ask people what they saw. Share war stories."

"The end of totality is a time for celebration," says Jay Anderson. "In the wake of such a powerful shared experience, conversations become animated and casual acquaintances tend to become lifelong friends."

Charles Piazzi Smyth, astronomer royal of Scotland, saw his first total eclipse in 1851, and recognized its ability to overwhelm an observer, distracting him from his carefully planned research. "Although it is not impossible but that some frigid man of metal nerve may be found capable of resisting the temptation," he wrote, "yet certain it is that no man of ordinary feelings and human heart and soul can withstand it."[13]

Confessions of Eclipse Junkies

What happens if you are clouded out, as happens even with good planning about one out of every six times? Unlike a rocket countdown, which can be "T minus 30 minutes and holding" while the

Chasing the Eclipse of 16 February 1980

by Fred Espenak

In February of 1980, I traveled to Kenya with a group of nearly a hundred people to witness a total eclipse of the Sun. Once in East Africa, our group was divided up for local transportation and wildlife viewing via VW minibuses.

Although the sky was clear on eclipse morning, weather changes quickly in Kenya. My minibus group decided to go into the bush where we would have plenty of maneuverability if conditions took a turn for the worse. The rest of the expeditioners elected to remain at our deluxe hotel, near the eclipse centerline. After breakfast, the four of us left the hotel with our driver Ali, who took us east to a sisal plantation. There we found a soccer field with a good view to the west to watch the Moon's shadow approach.

We set up our telescopes and observed first contact about 10 A.M. But as the eclipse progressed, clouds began to form across the entire sky. Presumably they were caused by the eclipse itself—the atmosphere cooling because of reduced sunlight. By 11 A.M., the cloud cover had increased to 75 percent. With only twenty minutes left before second contact, we began to consider moving. But moving would require us to leave much of our equipment behind, so we were reluctant. Besides, it was not at all clear which way we should run.

Seven minutes before totality, the sky was 95 percent covered by clouds. We were losing sight of the Sun for a minute or two at a time. We studied the valley below, timing the speed of slowly moving patches of sunlight across its floor. A hole in the cloud cover was drifting our way, but it would not reach us in time.

With just four minutes left, we decided to run! Grabbing only one camera with a long lens, I jumped into the front seat of our bus to direct Ali. Margie Milne brought the tape recorder and quartz clock, while husband David dragged one telescope into the back of the bus. In a matter of seconds we were off as the locals cheered us on. David, trapped under his telescope, made several attempts to close the bus door and end the trail of soda bottles, lens caps, and empty film boxes we were leaving in our wake.

---

clouds clear, an eclipse countdown is inexorable, and it is not always possible to race to another location where there is a break in the clouds.

"My first eclipse was in Maine in the summer of 1963," Joe Hollweg recalls. "I was working in New York City, but I made a trip to Maine just for the eclipse. The sky was partly cloudy, with lots of small cottony fair-weather clouds. What a disappointment when one of those little puffs moved over the Sun just at the start of totality! I

We concentrated on finding a hole in the clouds. We attempted to go west, but found the road doubled back the way we had come—and into the clouds. The air filled with the sound of grinding gears as Ali spun the bus in a frantic U-turn. Two minutes left! Back on the main dirt road, we headed south toward another hole in the clouds. As we caught up to the edge of the hole, the thin sliver of the nearly eclipsed Sun moved into the clearing. Onward. We wanted to get to the other edge of the hole for the beginning of totality, so the Sun would stay in view.

As the bus clattered down the dirt road at 100 kilometers per hour, I watched the diamond ring form while hanging out the front window. After timing the contact, we traveled another 100 meters and screeched to a stop. I leapt out the front door, but panic exploded in the back of the bus. The rear door wouldn't open. My companions dove over telescopes and seats to emerge from the front door. In the meantime, Ali rushed around to unlock the door and remove our equipment. By now, I had my tripod set up and was recording exposure information as I photographed the corona. Margie set up her telephoto and was trying to focus it when she suddenly screamed, "There's no camera attached to this lens!" David was busy viewing and photographing the spectacle with his refractor. Betty Pynn scrambled for binoculars. Ali simply smiled as he watched his strange tourists.

The corona was very bright, with a large streamer to the northwest. Even through the clouds, lots of structure was visible during this sunspot-maximum eclipse. Through my camera viewfinder, I spotted a large prominence. To the southeast, the sky was bright orange as I looked outward past the edge of the Moon's shadow.

Time passed quickly, as did our hole in the clouds. Twenty seconds before third contact, the corona was lost from view as clouds covered the Sun. I grabbed the tape recorder and tripod and went sprinting down the road after the hole, with David and Margie in close pursuit. We caught up to the hole and watched third contact on the run.

As we strolled back to our minibus, I realized that until now I had never fully appreciated the term "eclipse chaser."

P.S. Our comrades back at the hotel enjoyed a clearer view of the eclipse from poolside, but they missed a most memorable adventure!

got to see Baily's Beads, and that was all. A few hundred yards away people were in the clear, and they were cheering with excitement. However, I do remember seeing the Moon's shadow racing across those puffy clouds, both at the start of the eclipse and at the end. The motion of the shadow seemed incredibly rapid, perhaps because the clouds weren't very high. That was the first time I had some sense of how fast 1,000 miles per hour really is."

There were high hopes for the total eclipse of August 19, 1887, but

Clouded out. Glum observers at the August 9, 1896 eclipse in Lapland see only the twilight glow on the horizon in this painting by Lord Hampton. [Annie S. D. Maunder and E. Walter Maunder: *The Heavens and Their Story*]

clouds spoiled the view along almost the entire path from Germany through Russia to Japan. Someone posted a public notice in Berlin stating that, on account of the weather, the eclipse had been postponed to another day.[14]

Bad luck plagued Rebecca Joslin in her eclipse chasing. An amateur astronomer, she started chasing eclipses as a college student in 1905. Going to Spain in that era meant weeks of travel. The weather was perfect until about 10 minutes before totality, when a single cloud appeared, blotted out the Sun for the entire 3½ minutes of totality, and then moved on. She tried again in August 1914, sailing first for England, with reservations on a ship that would take her to a viewing site in Norway. While she was en route to England, World War I broke out, her England-to-Norway cruise was canceled, and she watched a partial eclipse from England under absolutely clear skies.

It was not until January 24, 1925, twenty years after her first attempt, that Joslin finally saw a total eclipse. This time the eclipse came almost to her. She had only to cross from Massachusetts into Connecticut. She had previously been wiped out by one cloud and one world war. Nature didn't make it easy on her this time either. The temperature at her site was -2°F (-19°C). One brief eclipse out of three. She gathered her travels and experiences into a book called *Chasing Eclipses*. Because of her keen eye for people, places, history, and irony, she salvaged quite a bit from misfortune.

All the veterans agree: Plan your eclipse trip so that you see things, go places, and meet people that will shine in your memory even if the corona does not.

Crucial to the enjoyment of a solar eclipse is eye safety. You wouldn't stare directly at the Sun during a normal day, so you shouldn't stare at the Sun when it is partially eclipsed without proper eye protection (described in the next chapter). However, when the Sun is totally eclipsed—no portion of its disk is showing—it is perfectly safe to look at the Sun without any eye protection. Jay Pasachoff is irritated by governments and news media in foreign lands and even in this country that mislead the public by implying that the Sun gives off "special rays" at the time of an eclipse. Instead of teaching simple safety procedures, they frighten people, thereby depriving them of a rare and magnificent sight.

Alan Fiala recalls the 1980 eclipse in India where pregnant women were instructed to remain indoors during the eclipse. For the 1992 eclipse, Roger Tuthill rented a jumbo jet in Brazil to meet and chase the shadow over the Atlantic, using the speed of the aircraft to expand three minutes of totality into six. To serve the fifty passengers aboard the DC-10, the airline assigned twelve flight attendants. There were plenty of windows for them to watch the eclipse, but almost all the stewardesses refused to look. They feared that if they saw the Sun in eclipse, they could never become pregnant.

The pilot, however, was not afraid to look. He was wildly enthusiastic. Even after the Moon's shadow outran the jet and totality was over, the pilot pressed on eastward toward Africa because he was so fascinated by the fast-moving pillar of darkness ahead of him. "It's even more exciting than an engine fire," he explained.

George Lovi, Alan Fiala, and others remember the 1983 eclipse in Indonesia. Before the eclipse, the streets were teeming with people. On eclipse morning, however, there was scarcely anyone outdoors. Fire sirens wailed like an air raid warning for people to take cover. Schoolchildren, on government orders, were kept indoors, forbidden to see the eclipse, except perhaps on television. A stunning natural event that comes to you, if you are lucky, once in five lifetimes, was denied to them.

On Java for that 1983 eclipse, Jay Anderson recalls soldiers patrolling the streets to discourage unauthorized observers, although the citizens were so frightened by government warnings that almost all stayed indoors. As the Sun was gradually disappearing, the soldiers near him began to be caught up in the excitement. Anderson offered them a view through his telescope, but the warnings they had received overtook them and they refused. However, several observers gave the officers a few minutes of careful explanation, after which the officers nervously took a peek. "After their first look all apprehension disappeared, and they participated in the event as fully as we did. Ten minutes before totality, the officer in charge was interrupted by a call on his walkie-talkie from his superior at headquarters:

"'How are things going out there?'

The Personalities of Eclipses

by Stephen J. Edberg

Weather, geography, and companionship profoundly affect the experience of an eclipse. But each eclipse has intrinsic differences from all others that lure eclipse veterans. They use these differences in planning their observations.

One factor is the magnitude of the total eclipse, the degree to which the disk of the Moon more than covers the disk of the Sun. When the angular size of the Moon is great, the Moon at mid-totality will mask not only the Sun's photosphere but also its chromosphere and lowermost corona. Except at the beginning and end of totality (or near the eclipse path limits), the stunning fluorescent pink of the prominences will be hidden from view, unless an absolutely gigantic prominence happens to be present on the Sun's limb.

However, because the relatively bright lower corona is blocked from view, a large-magnitude eclipse is the best time to observe the full extent and detail in the corona. Because the Moon appears larger, its shadow is wider. Thus the total eclipse lasts longer and the darkness is deeper, allowing the full majesty of the corona to shine through, as well as any planets and bright stars that happen to be above the horizon.

By contrast, in a smaller-magnitude total eclipse, the Moon's disk is not big enough to mask the lower corona, and prominences may be seen all the way around the disk of the Sun during most of totality. But this eclipse will be comparatively brief and the full extent of the corona may not be evident. A total eclipse with less obscuration often provides a better display of Baily's Beads and the Diamond Ring Effect. Because the Moon's disk is nearly the same apparent size as the Sun's disk, as totality nears, the crescent of the Sun is long and narrow, allowing Baily's Beads to glimmer like jewels on a necklace. When the Moon's apparent disk is large, the length of the Sun's crescent is greatly shortened as it narrows. There may be only one or two Baily's Beads.

As veteran observers plan for upcoming eclipses, they also take into consideration the sunspot cycle. At sunspot maximum, the corona is brighter, rounder, and larger. Prominences also tend to be more numerous. At

"'Okay. There are no problems.'

"'It's time to come in now. Collect your men and bring them back to the barracks. We are supposed to have all troops back before the eclipse begins.'

"'There are no problems here; everyone is safe; the tourists are all working on their equipment. Are you sure you want us in?—I see no problems.'

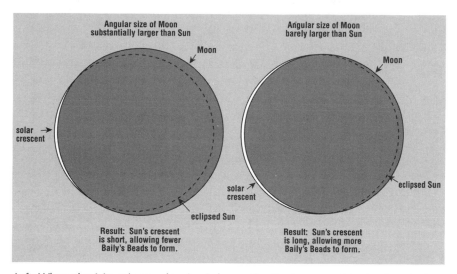

Left: When the Moon's angular size is large, the Sun's crescent is short, reducing the span of Baily's Beads. *Right*: When the Moon just barely covers the Sun, so that the disks appear nearly the same size, the Sun's crescent is much longer and the span of Baily's Beads is greater.

sunspot minimum, the corona is fainter and broader at the equator than at the poles. Brushlike coronal features projecting from the poles are more noticeable.

A third factor used by eclipse followers is the proximity of the shadow path to the Earth's equator, where the rotation of the Earth causes the ground speed of the Moon's shadow to be slowest, making the duration of totality longest. Eclipses with six to seven minutes of totality are almost always found between the Tropic of Cancer and the Tropic of Capricorn. The rotation of the Earth extends not just the period of totality but all aspects of the eclipse, so that, near the equator, the partial phases of the eclipse last longer. The last crescent of the Sun is covered more slowly, so Baily's Beads and the Diamond Ring Effect, while still brief, last a little longer, and the view of the prominences at the limb of the Sun is prolonged.

My Favorite Eclipse

by Ken Willcox

Every total eclipse is different—and wonderful—but if you see more than one, you probably have a favorite, for some special reason. My favorite was the eclipse of November 3, 1994, on the Altiplano in Bolivia.

The Altiplano, or Puno, is a wide flat desert high in the Andes Mountains of South America. It is 105 miles (169 kilometers) across, 311 miles (501 kilometers) long, and 9,800 to 13,200 feet above sea level. It was there we had come, 110 of us from all over the United States and Europe, to a small plot of land 12,516 feet (3,815 meters) above sea level, hoping for a spectacular view of a total eclipse of the Sun.

The Altiplano site was selected because it provided the best chance of clear skies along the eclipse path. But the price of clear skies was the major logistical problem of transporting 110 people safely and comfortably to a remote high-altitude viewing site in a foreign land.

On Wednesday, November 2, we boarded our own private train in La Paz, with a dining car and also a baggage car to accommodate the array of equipment brought to record three precious minutes of time. At 4:50 P.M., our chartered narrow-gauge train began its spiraling climb out of the trench in which La Paz lies, then turned south toward our destination 200 miles (320 kilometers) away, a spot on the centerline, 7½ miles (12 kilometers) south of Sevaruyo. In the middle of the night, using a global positioning system (GPS), Jim Zimbelman of the Smithsonian Institution and I navigated the train to a precise spot along the tracks selected two years earlier on a site-inspection trip.

As we approached our site about 4 A.M., we were met by 70 gun-toting soldiers provided by the Bolivian army. Under floodlights, before dawn, in the middle of nowhere high in the Andes, amateur and professional astronomers and Bolivian soldiers dragged equipment from the train and prepared to record an extraordinary celestial event. The soldiers set up a perimeter to keep out local people who might wander into the forest of telescopes that had suddenly sprouted in the desert.

"'Orders are to bring everyone in, and leave the tourists to themselves. Bring the men in.'

"'I'm sorry, I can't hear you. I'm having trouble with the radio. What did you say?'

"At this point the officer turned off the radio and put it back in his pocket. The entire troop stayed to watch the eclipse with us."

Even more memorable to Anderson were two dozen Indonesian students from a local college who were studying English had attached themselves to Anderson's group to practice their speaking skills. As

One hour after sunrise, the clouds began breaking up. Only high scattered cirrus remained at first contact at 7:19 A.M., and those too were vanishing. Approximately a half hour before totality, the wind began to increase from the east, rather than decrease as it usually does preceding a total eclipse of the Sun. The heat from the Sun falling on the Altiplano and its surrounding mountains was being extinguished by the increasing eclipse. The cooling was most rapid in the mountains and the heavier cold air rushed down the closer eastern slopes and onto the Altiplano.

A family of Aymara Indians, descendants of the Incas, had been invited to join us and were given solar filters to view the partial phases. This Bolivian family huddled around their father as totality approached. We kept glancing at them to see what their response to the disappearing Sun and the onset of totality would be. A few minutes before totality, the father made his three boys look down at the ground until the eclipse was over. The father feared the souls of the boys would be unalterably affected.

"Here it comes!" someone shouted. "Where?" "Over there," pointing to the northwest. "Oh yes! I see it!" "It's getting dark now . . . it's getting real dark now . . . it's getting really dark now . . . it's really getting . . . *oh my God!*" Cheers and gasps accompanied the beginning of totality. It took everything I had to keep my mind focused on the task at hand, and even that didn't work when the lady behind me broke down crying. Her husband wrapped his arms around her.

As totality ended, I shot a last few exposures blinded by my own tears, gave up, and turned around and photographed the couple behind me.

As totality began, astronomer Chris Halas of Lincoln, Massachusetts, felt as if her soul had been raptured into the presence of God, only to fall back to Earth at the end of totality. That comes closest to describing what it is like to experience a total eclipse of the Sun—except you don't want it to end, you don't want to come back to Earth.

The Aymara Indians were right. All our souls were unalterably affected by that eclipse. But it was—and is—a sublime experience, one that should be sought, not feared.

---

totality neared, the students became extremely nervous. To reassure them, Anderson and his fellow observers held hands with them during totality—the most magical moment in all the eclipses he has seen.

But less developed countries aren't the only places where superstition and misinformation reign supreme and deprive people of a rare gift of nature. The eclipse of 1970 was the first and most impressive of many George Lovi has seen. It was visible in the eastern United States and he was watching from Virginia Beach. The news media had trumpeted the event, but laid great emphasis on the dangers. Around him were thousands of people, most of them casual observers, watching as the crescent Sun thinned and vanished. Instantly a cry went up, spreading from group to group: "Look away! Look away!" Lovi knows of other curious people who traveled substantial distances to reach Virginia Beach and then, fearful, stayed in their motel rooms and watched the event on television.

Dennis di Cicco was in a cornfield in North Carolina for the 1970 eclipse and recalls that in the midst of totality, a car came up the road with its lights on and drove right by without stopping, its passengers oblivious or impervious to the wonder above them. Di Cicco also remembers Australia in 1976, where government warnings that people should stay in their houses to avoid the dangers of the eclipse were so intense that he and other observers feared that local citizens or the police might stop them from watching the eclipse out of concern that they were hurting themselves.

Steve Edberg was in Winnipeg, Canada, for the eclipse of 1979. Students had to have notes from their parents to be allowed outside during the eclipse. All the others were confined to their classrooms, where the shades were drawn and the students watched the eclipse on television. Jay Pasachoff recalls that "one school in Winnipeg even asked for permission to ignore fire alarms if any sounded during the eclipse, lest the students rush outside and be blinded." On a more rational note, Jay Anderson remembers that many parents kept their children home from school on eclipse day and took the day off themselves so that they could witness the eclipse as a family.

Sometimes superstitions and misinformation about eclipses can be tragic. In the aftermath of the 1998 eclipse in the Caribbean, an Associated Press wire story[15] reported:

> They were frightened by an eclipse—and apparently, they were frightened to death.
>
> Four members of a Haitian family have been found dead in their homes. Officials say the four may have been killed by an overdose of sleeping pills they took to alleviate their fears of last Thursday's eclipse. They also may have suffocated. They'd plugged all the openings to their home with rags, to keep the Sun out.
>
> Radio broadcasts in Haiti say another young girl suffocated Thursday in a home that had been sealed. Thousands of Haitians were afraid that the eclipse would blind them or kill them.

Stages of a Total Eclipse

First Contact   The Moon begins to cover the western limb of the Sun.

Crescent Sun   Over a period of about an hour, the Moon obscures more and more of the Sun, as if eating away at a cookie. The Sun appears as a narrower and narrower crescent.

Light and Color Changes   About 15 minutes before totality, when 80 percent of the Sun is covered, the light level begins to fall noticeably, and then it falls with increasing rapidity and the landscape takes on a metallic gray-blue hue.

Gathering Darkness on the Western Horizon   About 15 minutes before totality, the shadow cast by the Moon becomes visible on the western horizon as if it were a giant but silent thunderstorm.

Animal, Plant, and Human Behavior   As the level of sunlight falls, animals may become anxious or behave as if nightfall has come. Some plants close up. Notice how the people around you are affected.

Temperature   As the sunlight fades, the temperature may drop perceptibly.

Shadow Bands   A few minutes before totality, ripples of light may flow across the ground and walls as the Earth's turbulent atmosphere refracts the last rays of sunlight.

Corona   Up to one minute before totality, the corona begins to emerge.

Baily's Beads   About 10 seconds before totality, the Moon has covered the entire face of the Sun except for a few rays of sunlight passing through deep valleys at the Moon's limb, creating the effect of jewels on a necklace.

Shadow Approaching   While all this is happening, the Moon's dark shadow in the west has been growing. Now it rushes forward and envelops you.

Diamond Ring Effect   The Moon covers all but one of Baily's Beads, and that bead vanishes, often as if sucked into an abyss.

Second Contact   Totality begins. The Sun's photosphere is completely covered by the Moon.

Prominences and the Chromosphere   For a few seconds after totality begins, the Moon has not yet covered the lower atmosphere of the Sun and a thin strip of the vibrant red chromosphere is visible at the Sun's eastern limb. Stretching above the chromosphere and into the corona are the vivid red prominences.

Corona Extent and Shape   The corona and prominences vary with each eclipse. How far (in solar diameters) does the corona extend? Is it round or is it broader at the Sun's equator? Does it have the appearance of short bristles at the poles? Look for loops, arcs, and plumes that trace solar magnetic fields.

Landscape Darkness and Horizon Color   Each eclipse, depending mostly on the Moon's angular size, creates its own level of darkness. Beyond the Moon's shadow, at the far horizon all around you, the Sun is shining and the sky has twilight orange and yellow colors.

Temperature   Is it cooler still? A temperature drop of about 4°F (2°C) is typical.

Stages of a Total Eclipse (*continued*)

Animal, Plant, and Human Reactions   What animal noises can you hear? How do *you* feel?

Planets and Stars Visible   Venus and Mercury are often visible near the eclipsed Sun, and other bright planets and stars may also be visible, depending on their positions and the Sun's altitude above the horizon.

End of Totality Approaching   The western edge of the corona begins to brighten and the reddish prominences and chromosphere appear.

Third Contact   One bright dot of the Sun's photosphere returns to view. Totality is over. The stages of the eclipse repeat themselves in the opposite order.

Diamond Ring Effect

Shadow Rushes Eastward

Baily's Beads

Corona Fades

Shadow Bands

Crescent Sun

Recovery of Nature

Partial Phase

Fourth Contact   The Moon no longer covers any part of the Sun. The eclipse is over.

The government declared a national holiday, in hopes of preventing panic. Police ordered pedestrians off the street during the eclipse, yelling, "Go home! It's dangerous to be out!"

The great astronomy popularizer Camille Flammarion, quoting François Arago, tells of the eclipse of May 22, 1724, visible from Paris. Supposedly, a marquis and some aristocratic lady friends were invited to observe the eclipse at the Paris Observatory. However, the ladies fussed so long with their gowns and hairstyles that the party arrived late, a few minutes after totality had ended. "Never mind, ladies," said the marquis, "we can go in just the same. M. Cassini [the observatory director] is a great friend of mine and he will be delighted to repeat the eclipse for you."[16]

11

Observing Safely

Shadow and sun—so too our lives are made—
Here learn how great the sun, how small the shade!
               —Richard Le Gallienne (1920?)

WARNING:

Permanent eye damage can result from looking at the disk of the Sun directly, or through a camera viewfinder, or with binoculars or a telescope even when only a thin crescent of the Sun or Baily's Beads remain. The 1 percent of the Sun's surface still visible is about 4,000 times brighter than the full moon. Staring at the Sun under such circumstances is like using a magnifying glass to focus sunlight onto tinder. The retina is delicate and irreplaceable. There is little or nothing a retinal surgeon will be able to do to help you. Never look at the Sun outside of the total phase of an eclipse unless you have adequate protection.

Once the Sun is entirely eclipsed, however, its bright surface is hidden from view and it is completely safe to look directly at the totally eclipsed Sun without any filters. In fact, it is one of the greatest sights in nature.

There are five basic ways to observe the partial phases of a solar eclipse without damage to your eyes.

The Pinhole Projection Method

One safe way of enjoying the Sun during a partial eclipse—or anytime—is a "pinhole camera," which allows you to view a *projected*

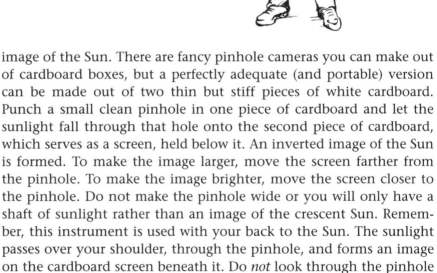

Pinhole camera made from two pieces of cardboard. Sunlight falls through the pinhole and forms an inverted image of the Sun on the screen. [Drawing by Sheri Flournoy, University of Tennessee]

image of the Sun. There are fancy pinhole cameras you can make out of cardboard boxes, but a perfectly adequate (and portable) version can be made out of two thin but stiff pieces of white cardboard. Punch a small clean pinhole in one piece of cardboard and let the sunlight fall through that hole onto the second piece of cardboard, which serves as a screen, held below it. An inverted image of the Sun is formed. To make the image larger, move the screen farther from the pinhole. To make the image brighter, move the screen closer to the pinhole. Do not make the pinhole wide or you will only have a shaft of sunlight rather than an image of the crescent Sun. Remember, this instrument is used with your back to the Sun. The sunlight passes over your shoulder, through the pinhole, and forms an image on the cardboard screen beneath it. Do *not* look through the pinhole at the Sun.

Mylar Sun Filters

A second technique for viewing the Sun safely is by looking directly at the Sun through an aluminized Mylar filter. Advertisements for such filters may be found in popular astronomy magazines. Beware, though, that Mylar, a plastic, comes in various thicknesses and with various coatings. You need a metal coating to save your eyesight and you need to examine the Mylar for small holes that could allow unfiltered sunlight to reach your eyes and damage them. A good solar filter will allow you to look comfortably at the filament of a high-intensity lamp.

When using any filter, however, do not stare for long periods at the Sun. Look through the filter briefly and then look away. In this way, a tiny hole that you miss is not likely to cause you any harm. You know from your ignorant childhood days that it is possible to glance at the Sun and immediately look away without damaging your eyes. Just remember that your eyes can be damaged without you feeling any pain.

Welder's Goggles

Welder's goggles or the filters for welder's goggles with a rating of 14 or higher are safe to use for looking directly at the Sun. They are also relatively inexpensive.

Camera and Telescope Solar Filters

Many telescope and camera companies provide metal-coated filters that are safe for viewing the Sun. They are more expensive than common Mylar, but observers generally like them better because they are available in various colors, such as a chromium filter through which the Sun looks orange. Through aluminized Mylar, the Sun is blue-gray. As with the Mylar, you can look directly at the Sun through these filters.

Caution: Do not confuse these filters, which are designed to fit over the lens of a camera or the aperture of a telescope, with a so-called solar *eyepiece* for a telescope. Solar eyepieces are still sometimes sold with small amateur telescopes. *They are not safe* because of their tendency to absorb heat and crack, allowing the sunlight concentrated by the telescope's full aperture to enter your eye.

Fully Exposed and Developed Black-and-White Film

You can make your own filter out of black-and-white film, but *only* true black-and-white film (such as Kodak Tri-X or Pan-X). Open up a roll of black-and-white film and expose it to the Sun for a minute. Have it developed to provide you with negatives. Use the negatives (black in color) for your filter. It is best to use two layers. With this filter, you can look directly at the Sun with safety.

Remember, however, that if you are planning to use black-and-white film as a solar filter, you need to prepare it at least several days in advance.

Caution: Do not use color film or chromogenic black-and-white

Eye Damage from a Solar Eclipse

by Lucian V. Del Priore, M.D., Ph.D.

The dangers of direct eclipse viewing have not always been appreciated, despite Socrates' early warning that an eclipse should be viewed only indirectly through its reflection on the surface of water. A partial eclipse in 1962 produced 52 cases of eye damage in Hawaii, and a total eclipse along the eastern seaboard of the United States produced 145 cases in 1970. As many as half of those affected never fully recovered their eyesight.

There is nothing mysterious about the optical hazards of eclipse viewing. No evil spirits are released from the Sun during a solar eclipse, and there is no scientific reason for running indoors to avoid "the harmful humors of the Sun." Eye damage from eclipse viewing is simply one form of light-induced ocular damage, and similar damage can be produced by viewing any bright light under the right (or should I say the wrong!) conditions.

Light enters the eye through the cornea and is focused on the retina by the optical system in the front of the eye. Any light that is not absorbed by the retina is absorbed by a black layer of tissue under the retina called the retinal pigment epithelium. The retina is the human body's video camera; it contains nerve cells that detect light and send the electrical signal for vision to the brain. Without it, we cannot see. Most of the retina is devoted to giving us low-resolution side vision. Fine detail reading vision is contained in a small area in the center of the retina called the fovea. People who damage their fovea are unable to read, sew, or drive, even though this small area measures only 1/100th of an inch across, and is less than 1/10,000th of the entire retinal area! Unfortunately, this is the precise area that is damaged if we stare directly at the surface of the Sun. Damage has been reported with less than one minute of viewing. The image of the Sun projected onto the retina is about 1/150th of an inch in size, and this is large enough to seriously damage most of the fovea.

Why is sunlight damaging to a structure that is designed to detect light? Bright sunlight focused on the retina is capable of producing a thermal burn, mainly from the absorption of infrared and visible radiation. The

film (which is actually a color film). Developed color film, no matter how dark, contains only colored dyes, which do not protect your vision. It is the metallic silver that remains in black-and-white film after development that makes it a safe solar filter.

Eye Suicide

Standard or polaroid sunglasses are not solar filters. They may afford some eye relief if you are outside on a bright day, but you would never think of using them to stare at the Sun. So you cannot

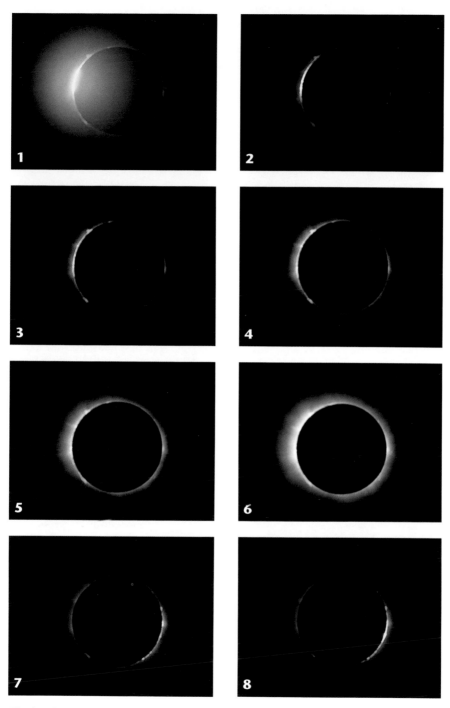

The beginning and end of totality at the February 26, 1979 eclipse, showing the Diamond Ring, prominences, the inner corona, and Baily's Beads. [Celestron 8-inch, 2000 mm focal length, f/10, shutter speeds: (1) 1/250, (2) 1/250, (3) 1/250, (4) 1/250, (5) 1/250, (6) 1/60, (7) 1/500, (8) 1/500, with ISO 400 film. © 1979 Ken Willcox]

Partial solar eclipses are especially beautiful when they occur during sunrise or sunset. December 13, 1974 from Phoenix, Arizona. [48 mm refractor (no solar filter), 340 mm focal length, f/7, 1/30 second, ISO 25 film. © 1974 Ernie Piini]

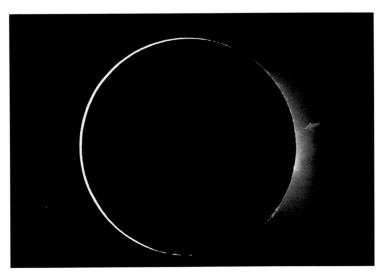

During the annular eclipse of May 30, 1984, Dennis di Cicco used a 4.0 neutral density filter to cover half the camera's film plane. By orienting the filter to cover the Sun's bright crescent, he could also capture Baily's Beads and prominences along the Sun's opposite limb. [Jagers 4⅛-inch refractor, 1570 mm focal length, f/15, 1/500 second, ISO 64 film. © 1984 Dennis di Cicco]

Fred Espenak operates his cameras during the total eclipse of November 3, 1994 from La Lava, Bolivia [Nikon 8008, 16 mm fisheye lens, f/5.6, 4 seconds, ISO 100 film. © 1994 Fred Espenak and Ken Bertin]

The solar corona of November 3, 1994 from central Bolivia [3.5-inch Questar, 1440 mm focal length, f/16, 1 second, ISO 400 film. © 1994 Ken Willcox]

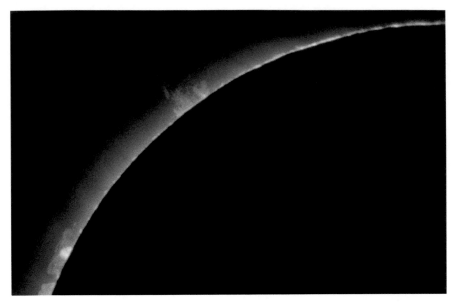

Ruby red prominences rise above the Moon's limb during the total solar eclipse of October 24, 1995 from Mandawa, India [Quantum 6-inch Maksutov, 2900 mm focal length, f/9, 1/30 second, ISO 64 film. © 1995 Jacques Guertin]

Two snow-capped volcanoes and a lake provide an impressive foreground during totality. Lake Chungara, Chile on November 3, 1994. [Nikon F4S, 24 mm lens, f/2.8, 1/4 second, ISO 50 film. © 1994 Serge Brunier]

Using five eclipse photos taken by Dennis di Cicco and Gary Emerson, Steve Albers digitally combined and enhanced the images using a custom computer program and 500 hours of hard work. His composite shows a wealth of details and fine structure in the corona during the total solar eclipse of July 11, 1991.

A composite image shows an entire eclipse at one glance. Sixty-three individual photos of the total solar eclipse of March 7, 1970 were scanned into a computer to make the final composition [60 mm refractor, f/15, 1/60 second, ISO 64 film; totality: 1 second, ISO 160 film. © 1998 Fred Espenak]

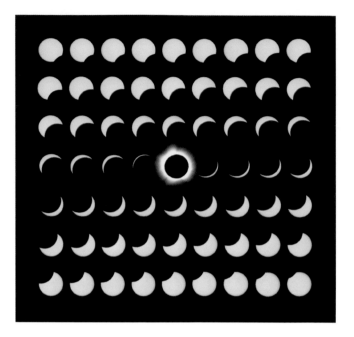

The eerie twilight of totality silhouettes astronomers as they quickly make their measurements during the eclipse of February 16, 1980, near Hyderabad, India. [Nikon FE, 24 mm, f/2.8, 1/4 second, ISO 200 film. © 1980 Jay M. Pasachoff]

Johnny Horne captured the moon's disk surrounded by a diamond necklace during the beaded annular eclipse of May 30, 1994 from Pendleton, South Carolina. [3-inch Jaegers refractor, no solar filter, 1140 mm focal length, f/15, 1/500 second, ISO 64 film. © 1984 Johnny Horne]

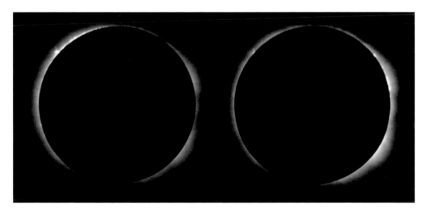

The Moon's motion with respect to the Sun is revealed in this pair of images taken at the beginning (*left*) and the end (*right*) of totality on February 26, 1998 from Oranjestad, Aruba. A large prominence and the inner corona are easily seen at the 10 o'clock position in the left image. Three minutes later, they are almost covered as the inner corona and chromosphere are now revealed at the 2 o'clock position in the right photo. [3.5-inch Questar, 1440 mm focal length, f/16, 1/125 second, ISO 400 film. © 1998 Ken Willcox]

Using a computer, Fred Espenak combined 20 photographs to produce this image of the total eclipse of February 26, 1998 from Oranjestad, Aruba. This special processing captures both bright and faint details and closely resembles the corona's appearance to the naked eye. [© 1998 Fred Espenak]

The Diamond Ring Effect at third contact signals the end of totality. Total solar eclipse of February 26, 1998 from Oranjestad, Aruba [80 mm fluorite refractor, 640 mm focal length, f/8, 1/125 second, ISO 100 film. © 1998 Fred Espenak]

absorption of this light raises the temperature and literally fries the delicate ocular tissue. There is no mystery here, as every schoolchild knows that sunlight focused through a magnifying glass can cause a piece of paper to burst into flames. Yet Sun viewing seldom produces a thermal burn; all but the most intoxicated viewer would surely turn away before this occurs.

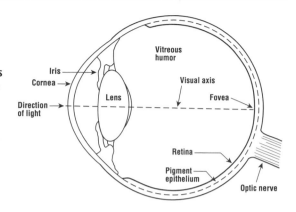

Cross section of the eye. [Drawing by Josie Herr]

Instead, most cases of eclipse blindness are related to photochemically induced retinal damage, which occurs at modest light levels that produce no burn and no pain. Two types of light lesions are recognized clinically and experimentally, and both are probably responsible for the damage observed after eclipse viewing. Blue light (400–500 nanometers) damages the retinal pigment epithelium and leads to secondary changes in the retina, while near ultraviolet light (340–400 nanometers) is absorbed by and directly damages the light-sensitive cells in the outer retina.

Viewing a partial eclipse recklessly is not the only way to produce light-induced retinal damage. Sungazing is a well-known cause of retinal damage even in the absence of an eclipse. Numerous cases of blindness have been reported in sunbathers, in military personnel on antiaircraft duty, and in religious followers who sungaze during rituals and pilgrimages. The Sun is not even required to produce light damage; other types of bright lights, including lasers and welder's arcs, will have the same effect. The common thread here is clear: direct viewing of bright lights can damage the retina regardless of the source. A solar eclipse merely increases the number of potential victims, and brings the problem to public attention.

use sunglasses, even crossed polaroids, to stare at the Sun during the partial phases of an eclipse. They provide little or no eye protection for this purpose.

Observing with Binoculars

Binoculars were astronomy writer George Lovi's favorite instrument for observing total eclipses. Any size will do. He used 7 × 50 (magnification of 7 times with 50-millimeter [2-inch] objective lenses). "Even the best photographs do not do justice to the detail and color of the

Sun in eclipse, and especially the very fine structure of the corona, with its exceedingly delicate contrasts that no film can capture the way the eye can." The people who did the best job of capturing the true appearance of the eclipsed Sun, he felt, were the nineteenth-century artists who photographed totality with their eyes and minds and developed their memories on canvas.

For people who plan to use binoculars on an eclipse, Lovi cautioned common sense. Totality can and should be observed without a filter, whether with the eyes alone or with binoculars or telescopes. But the partial phases of the eclipse, right up through the Diamond Ring Effect, must be observed with filters over the objective lenses of the binoculars. Only when the Diamond Ring has faded is it safe to remove the filter. And it is crucial to return to filtered viewing as totality is ending and the western edge of the Moon's silhouette begins to brighten. After all, binoculars are really two small telescopes mounted side by side. If observing the Sun outside of eclipse totality without a filter is quickly damaging to the unaided eyes, it is far quicker and even more damaging to look at even a sliver of the uneclipsed Sun with binoculars that lack a filter.

Observing with a Telescope

Some observers, including astronomy historian Ruth Freitag, prefer to watch the progress of the eclipse through a small portable telescope, which offers stability and is much less tiring to use for extended periods than binoculars. A telescope also provides more detail at higher powers, if this is desired. The solar filter is removed at totality. When she wants to see a wider view of the corona, she switches to the finder scope.

A Final Thought

Just remember, Lovi said, "don't try to do too much. Look at the eclipse visually. Don't be so busy operating a camera that you don't see the eclipse. And don't set off for the eclipse so burdened down by baggage and equipment that you are tired and stressed and too nervous to enjoy the event."

Astronomer Isabel Martin Lewis also warned of the dangers of too many things to do: "A noted astronomer who had been on a number of eclipse expeditions once remarked that he had never *seen* a total solar eclipse."[1]

12

Eclipse Photography

Each eclipse has at least one phenomenon that makes it special.
—Stephen J. Edberg (1990)

How do you capture the amazing spectacle of a total eclipse with a camera? Photographing an eclipse really isn't very difficult. It doesn't even take a lot of fancy or expensive equipment. You can take a snapshot of an eclipse with a 35 mm camera loaded with fast film (ISO 400 or faster) if you can hold the camera steady or place it on a tripod.

The first step in eclipse photography is to decide what kind of pictures you want. Are you partial to scenes with people and trees in the foreground and a small but distinct eclipsed Sun overhead? Or do you prefer a close-up in which the flaring corona or vibrant prominences of the eclipsed Sun fill the frame? Your decision will determine what kind of equipment you need. Look at the photographs and captions throughout this book. They illustrate some of what can be done with a range of cameras, lenses, and telescopes, and of film and exposures.

New technologies in film, cameras, and electronics are making eclipse photography easier than ever before. Even beginners can take great eclipse photos with some careful planning. *Planning* is the key. The day of the eclipse is *not* the time to try out a new tripod or lens. You need to be completely familiar with your camera and equipment, and you need to *rehearse* with them weeks before the eclipse. A total eclipse grants you only a few precious minutes, and everything must work perfectly. Nature does not provide instant replays.

The Right Film

Color film comes in two basic types: print film (negatives) and slide film (transparencies).[1] An advantage of print film is that it is very forgiving about exposure times. You can be off as many as three f-stops (or shutter speeds) from the best exposure and still produce an acceptable photograph. Slide film is less forgiving. If your exposure is off by more than one f-stop, your photograph is usually ruined. Slides are also less convenient to view than prints.

Slide film does have advantages, however. It requires just one processing step, which helps to retain accurate color. Slides can also be projected on a screen for large audiences.[2]

If you were allowed only one roll of film to photograph an eclipse, it is hard to go wrong with the latest ISO 400 film from Agfa, Fuji, Kodak, and Konica. The ISO rating is a measure of the film's sensitivity: how fast it responds to light. Films with an ISO of 100 or less are "slow" and work best in bright light. Films with an ISO of 400 or more are "fast" and can work in dimmer light.

There is a trade-off between film speed (ISO rating) and resolution. As film speed increases, so does the graininess of the film. Fortunately, film technology is advancing by leaps and bounds. Today's ISO 400 films are finer grained than the best ISO 100 films were only a decade ago.

A greater threat to image sharpness, especially among beginners, is camera vibration caused by wind, flimsy tripods, and nervous photographers. ISO 400 films allow you to use faster shutter speeds, which can help minimize these bad vibrations.[3]

In general, films that work well for everyday photography also work well for shooting eclipses.

The Right Solar Filters

When viewing or photographing the partial phases of any solar eclipse, you must always use a solar filter. A solar filter is also needed for observing all phases of an annular eclipse, when the disk of the Moon does not block the entire face of the Sun. Even if 99 percent of the Sun is covered, the remaining crescent or ring is dangerously bright. It is like looking at a welder's torch; it will painlessly burn your eyes. Failure to use a solar filter can result in serious eye damage or permanent blindness. *Do not look directly at the Sun without proper eye protection!*

During totality, however, when the disk (photosphere) of the Sun is fully covered by the Moon, it is completely safe to look at this

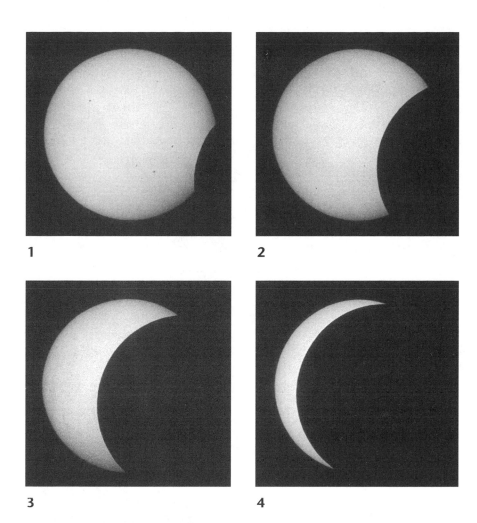

1

2

3

4

5

The disappearing Sun: partial phases approaching totality, February 26, 1979 [f/5, exposures of 1/30, 1/15, 1/15, 1/15, and 1/60 second, with ISO 200 film. © 1979 Ken Willcox]

phase of the eclipse without any solar filter. In fact, you *must* remove the solar filter during totality or you will not be able to see or photograph the exquisite solar prominences and corona.

Solar filters for telescopes and cameras are usually made of metal-coated glass to provide the highest resolution, although aluminized Mylar can also be used. These filters vary in the wavelength (color) of light they transmit. Aluminized Mylar filters show a blue-gray Sun, while the more expensive metal-coated glass filters transmit a more realistic orange Sun. Materials and techniques that should *not* be used for solar filters include exposed color film, stacked neutral density filters, smoked glass, and crossed polaroid filters.

There are three types of solar filters: eyepiece, off-axis, and full-aperture. Eyepiece filters, furnished with some small telescopes, *are not safe and should never be used for viewing the Sun.* The tremendous heat generated at the eyepiece can easily shatter or crack the filter, allowing the full intensity of the Sun's light to be magnified and focused on your retina. Throw eyepiece filters away.

The only safe solar filters for telescopes and cameras are full-aperture and off-axis filters, both of which fit over the objective

Diamond Ring Effect at February 16, 1980 total eclipse in Tsavo, Kenya. [Spiratone telephoto, 400 mm focal length, f/5.6, 1/125 second, with ISO 64 film. © 1980 Charles Simpson]

(front end) of the telescope or camera lens. A full-aperture solar filter is a cap with a solar filter mounted across its entire top. An off-axis solar filter is a cap with a hole off to one side into which the solar filter is mounted. Off-axis filters are cheaper than full-aperture filters because their filters are smaller.

If you use an off-axis solar filter with any catadioptric telephoto lens, the focus of the Sun will change significantly when you remove the filter to photograph totality. You must refocus. The optical field of a catadioptric system is not flat, and focusing with only the edge of the mirror is quite different from focusing with the whole mirror. In a full-aperture solar filter, the filter occupies the entire top of the cap and thus the focus is averaged over the entire surface of the mirror. No refocusing is needed when the filter is removed at the beginning of totality or when the filter is replaced at the end of totality. Full-aperture solar filters are preferred.[4]

If your telescope has a finder scope, be sure to place a small Mylar solar filter over its objective lens to protect your eyes and to keep the finder crosshairs from burning. If you don't have any Mylar, keep the lens cover on the finder scope.

Most telescope manufacturers and dealers offer glass solar filters for their products. Both types of solar filters are advertised in the major popular astronomy magazines (*Astronomy*, *Astronomy Now*, and *Sky & Telescope*). Solar filter retailers are also listed in Appendix C of this book.

The Right Cameras and Lenses

Close-ups of the eclipsed crescent Sun and detailed portraits of the solar corona require the use of a single-lens reflex (SLR) camera.[5] SLRs also feature interchangeable lenses, from extreme wide-angle to high-power telephoto. You can even remove the lens and hook the camera body up to a telescope so that the telescope provides the optics. The modern SLR is an electronic marvel that features auto-focus, programmed auto-exposure, and a built-in motor drive to advance the film. But older SLRs are also fine for eclipse photography. SLRs come in several film-size formats, but this chapter will concentrate on the 35 mm SLR.

No decision affects your eclipse photography more than the choice of a lens. Wide-angle lenses have focal lengths of 35 mm or less, while normal lenses fall within the range of 45 mm to 70 mm. Telephoto (high-magnification) lenses begin at 105 mm and go to 500 mm or 1000 mm. Zoom lenses provide adjustable magnification and new designs cover a wide range of focal lengths.[6] Additional magnification can be obtained with teleconverters. Teleconverters fit between the camera body and the lens to increase the lens's focal length by 1.4 to 2 times, thus increasing magnification.

Super Telephotos and Telescopes

The size of the Sun's image on your film is determined solely by your lens's focal length. For eclipse close-ups, a telephoto lens or telescope with a focal length of 500 mm or greater is recommended. A 500 mm focal length yields a solar image of 4.5 mm on a standard 35 mm negative. Allowing for one solar radius of corona on either side of the Sun, an image of the total eclipse covers about 9 mm on the film. You can calculate the Moon's or Sun's image diameter on your film (in millimeters) for any camera system by dividing the focal length (in millimeters) by 110.[7]

The least expensive 500 mm telephoto lens works like a catadioptric telescope. Catadioptric telescopes use a combination of mirrors and lenses to focus the light into your eye or onto film. Their folded light path allows a long focal length to fit within a short portable tube, making small catadioptric telescopes easily portable—ideal for most eclipse photography.[8]

Coupling a 2× teleconverter with a 500 mm lens will produce a 1000 mm focal length, which doubles the Sun's size to 9.1 mm. For the Sun and corona together, their image size increases to 18 mm.

The ideal focal length for telephoto photography of solar eclipses ranges from 500 to 2000 mm, depending on whether you are concentrating on the corona or on the prominences. Again allowing one solar radius of corona on either side of the Sun, you can calculate that lenses with focal lengths longer than about 1400 mm may not contain all the corona and focal lengths longer than 2500 mm may not capture the entire solar disk. The longer the lens, the more expensive and the less stable a telescope/telephoto system will be, which means that you will need a heavy-duty tripod or mount, adding to your expense and weight.

Improvements in lens technology have revived refracting telescopes as long lenses for eclipse photography. The new extra-low-dispersion (or apochromatic) lenses virtually eliminate the color halos around images that handicapped older refractors. Apochromatic refractors have wider objective lenses, yet shorter tubes, making them fast cameras. Apos require equatorial mounts and they cost more and are not as portable as catadioptric telescopes, but they are prized by advanced eclipse photographers for their image quality.[9]

Telescope Clock Drives and Polar Alignment

Because of the magnification required for eclipse close-ups and because your platform, the Earth, is rotating on its axis, a sturdy equatorial mount with clock drive is needed to take quality photographs with a telephoto lens or telescope with a focal length longer than about 1200 mm.

A clock drive slowly turns your telescope one rotation per day so that you can track the Sun, Moon, and stars from east to west as the Earth rotates. Without a clock drive, the Sun drifts through your camera's field at the rate of the Sun's apparent diameter every two minutes. The higher the magnification, the smaller your camera's field of view. Without a clock drive, you must frequently redirect your camera to center the Sun or it will drift to the edge or even out of the frame during the eclipse. A clock drive keeps the Sun centered so you can concentrate on watching and photographing totality. A clock drive also permits you to use longer focal lengths and take longer exposures without fear of blurring your photos due to the Earth's rotation.

Many modern telescopes are equipped with 12-volt battery-driven clock drives. A battery-driven clock drive frees you from worrying about AC access, voltages, frequencies, and plug sizes used in other countries. If your telescope uses DC power, bring two fresh 6-volt lantern batteries and wire them together in series. They will run your telescope for several hours.[10]

For your telescope to follow the Sun throughout the eclipse, the axis of the equatorial mount must be aligned with the celestial pole at the eclipse site. This task can be accomplished by arriving at your eclipse site the night before the eclipse and aligning on Polaris in the northern hemisphere or Sigma Octanis in the southern hemisphere. If the eclipse is south of the equator, make sure your clock drive has a reversible motor for use in the southern hemisphere.

More likely, however, you will have to set up your telescope just hours before the eclipse, with no opportunity for nighttime polar alignment. In this case, you will need a bubble angle finder and a good-quality magnetic compass. Bubble angle finders let you measure angles with respect to the horizontal and are available in most hardware stores.

First, level your equatorial mount and tripod using the bubble angle finder. Use your compass to align the azimuth of your polar axis on north. Since magnetic north can differ from celestial north by many degrees, you need to know the offset, or magnetic deviation, for your observing site and make that correction in azimuth. The National Geophysical Data Center Web site provides the magnetic deviation for the longitude and latitude of your observing site (www.ngdc.noaa.gov/cgi-bin/seg/gmag/fldsnth1.pl). You can also get this information from the nearest airport or from aviation or topographic charts.

Next, adjust the polar axis of the mount to the latitude of the observing site using your bubble angle finder. The polar axis should be set at an angle from the horizontal equal to your geographic latitude.[11]

Unless you have a clock drive, you must use relatively short exposures, because the rotation of the Earth causes the Sun to appear to drift westward ½° every two minutes. For your image to be in sharp focus, here is a formula for the longest exposure allowable with no clock drive:

$$\text{Exposure (seconds)} = 340 \text{ / focal length (millimeters)}[12]$$

For longer exposures, you need a clock drive to compensate for the Earth's rotation.

Camera Tripods

Flimsy tripods are the main reason that eclipse photographs come out fuzzy and blurred. Small portable tripods that are so nice for airline travel are not study enough to hold your camera and heavy telephoto lens steady for clear, sharp eclipse pictures. Because you will be touching your camera to adjust exposures, your tripod must also dampen any vibrations quickly.[13]

For greatest stability and to minimize vibrations, don't extend the tripod's legs more than halfway out and do not extend the center column. Try adjusting the height so that you can easily reach the camera controls while kneeling on the ground or sitting on a chair. Test your setup by tapping on the camera or tripod leg while viewing the Sun through your camera. Vibrations should be small and should damp out quickly.

You can further decrease vibrations to your camera by suspending some weight under the tripod. Put rocks or sand in a sack or in plastic zipper lock bags and hang them from the center of the tripod using string or duct tape.

Before you ever travel to an eclipse, make sure that your lens on its tripod can be pointed at the Sun's predicted altitude for the eclipse and verify that all controls work smoothly. You don't want to discover that your equipment becomes unstable when you try to view the crucial portion of the sky on eclipse day.

Cable Releases and Right Angle Finders

You cannot take crisp, high-quality close-ups of a total solar eclipse without a cable release specifically designed for your camera. Cable releases can fail unexpectedly, so have a spare.

A right angle finder is a little prism/eyepiece device that attaches to your SLR's viewfinder. It allows you to look through the viewfinder when your head is above the camera instead of behind it. Your camera will be pointed upward during the eclipse. Without a right angle finder, you might have to get down on your hands and knees or maybe lie on your back to look through your camera. A right angle finder allows you to comfortably center and focus the Sun's image. You will also be facing the top of your camera so you can see all the controls during totality.

Photographing the Partial Eclipse

You'll be photographing the partial phases of the eclipse with a solar filter over your lens. To succeed, you must determine well in advance of the eclipse the proper shutter speed and f-ratio for your particular solar filter and telephoto lens.

If your camera has a built-in spot meter that covers a smaller area than the Sun's image, you can simply meter on the Sun's disk through your solar filter and use that exposure throughout the partial phases.

If your camera does not have a spot meter (and most don't), you will need to run a simple exposure test. Set your equipment up on a sunny day. Load your camera with the same kind of film you will use for the eclipse. Use your solar filter to center the Sun carefully in your lens or telescope focus. For telephoto lenses, open the aperture to its widest setting. Shoot one exposure with every shutter speed you have from 1/15 through 1/1000 or 1/2000. Take notes so that you can identify the best exposure after your film is developed. Write down the best exposure and tape it to your tripod or the side of your solar filter. It should include the film speed, f-number, and shutter speed—for example, "Sun: ISO 400, f/8, 1/125." In this way the best exposure will be handy when you photograph the eclipse. The exposure doesn't change during the partial phases because the Sun's surface brightness remains the same throughout the eclipse.[14]

And now a warning. Your best exposure was determined on a sunny day. If the eclipse day has thin clouds or is hazy, you will need a longer exposure to compensate. A light haze may require an exposure one or two shutter speeds slower than normal, while thicker clouds could call for three or more shutter speeds slower. Try your planned exposure and several longer ones. Film is cheap and eclipses don't happen often.

Photographing the Total Eclipse

The brightness of the solar corona changes tremendously as you move out from the edge of the Sun's disk. The inner corona shines as brightly as the full moon, but the outer corona is over a hundred times fainter. The challenge is to capture both the brightest and faintest parts of the corona. Unfortunately, this variation in brightness is impossible to record in any one exposure because film just doesn't have the dynamic range of the human eye. (That's why you should look at totality with your eyes and not just with your camera. In real time, only your eyes can see the exquisite detail of this celestial event in all its glory.)

The good news is that you can photograph some aspect of the corona with almost any exposure you make. There is no one "correct" exposure. Nevertheless, here are some guidelines for where to start.

Several factors determine the length of time the camera shutter is open to get the proper exposure on your final photograph. A table accompanying this chapter provides recommended shutter speeds for various eclipse phenomena using a range of ISO film speeds and lens f-numbers. Each eclipse phenomenon (diamond ring, prominences, corona) has a different brightness value, and this value too must be considered in order to get the proper exposure of that aspect of the eclipse.[15]

Bracket your exposures on both sides of the ideal exposure and take several photographs at the same settings to help ensure success. If you use a film with an ISO too high, you may discover that your camera does not have a fast enough shutter speed to allow the proper exposure. Determine all of your camera settings before the eclipse so that you know in advance what film you will need with your equipment.[16]

Even if your SLR's exposure is completely automatic, or if you simply don't want to hassle with exposure settings, you can still get good pictures of the diamond ring and totality with your camera set on automatic exposure. Load your camera with ISO 400 color negative film and grab some shots on auto-exposure. You will probably over-

During the eclipse of February 16, 1980, the corona had a round shape, typical for sunspot maximum. [Bushnell spotting scope, 750 mm focal length, f/10, 2 seconds, with ISO 200 film. © 1980 Randy Attwood]

expose the inner corona and prominences, but you will still have some fine souvenirs of the event. Best of all, you can devote most of your time to watching the eclipse rather than fiddling with camera settings. Simplicity is especially recommended if you are a novice photographer or have never seen a total solar eclipse.

The Global Positioning System and Time Signals

The times for key phases of each eclipse are listed for any location along the eclipse path in NASA eclipse bulletins (Appendix G). To anticipate these events, you should set your watch for the exact time before the eclipse begins.

Coordinated Universal Time* is broadcast 24 hours a day over shortwave radio station WWV, Fort Collins, Colorado, and from Hawaii on WWVH at 5, 10 and 15 MHz. Radio station CHU in Ottawa, Canada, also broadcasts time signals at 3.330, 7.335, and 14.670 MHz. If you leave your radio on throughout the eclipse, you will always know the correct time without checking your watch.

Unfortunately, these time signals can seldom be picked up outside of North America and the central Pacific Ocean. A global positioning system (GPS) fills this void. A GPS receiver can provide not only correct time but also your precise location.

The GPS program was developed by the United States Department

---

*Coordinated Universal Time is equal to Greenwich Mean Time to within a second or two.

of Defense as a highly accurate way of determining geographic coordinates worldwide. The system uses a set of 24 satellites in orbits 12,545 miles (20,183 kilometers) high and with periods of 12 hours. At any time, eight to 12 of the GPS satellites are visible from any spot on Earth. A GPS radio receiver uses signals broadcast from the satellites to determine its three-dimensional position with respect to the satellites. This position is displayed as latitude, longitude, and elevation above sea level, along with the exact Universal Time. Military GPS receivers can determine positions to within inches. GPS receivers for civilian use have an accuracy of about 300 feet (100 meters). You can use them to assure yourself that you are standing within the path of totality.[17]

Tape Recorders

A small cassette tape recorder is very useful for recording your observations and reactions during the eclipse. Put in a fresh set of batteries before the eclipse begins. Fifteen minutes before totality, begin recording continuously. You will not only have a helpful and permanent record of your comments, but the tape will also capture the excitement of your companions.

Some advanced eclipse observers bring a second cassette recorder for which they have prerecorded a tape of instructions that they play back through headphones during the hectic minutes surrounding totality. The tape of instructions is synchronized with their watches and gives them audio cues for the time remaining before second or third contact, times to remove or replace solar filters, camera settings, and reminders to watch for various eclipse phenomena.

If you've never seen a total eclipse, such a tape may seem more trouble than it's worth. But the onset of totality can be so overwhelming that even experienced eclipse chasers lose track of what they intended to do. Consider an account from 1842:

> [T]he Captain of a French ship had beforehand arranged in the most careful way the observations to be made: but when the darkness came on, discipline of every kind failed, every person's attention being irresistibly attracted to the striking appearance of the moment, and some of the most critical observations were thus lost.[18]

Photographing Pinhole Crescents

Eclipses provide other phenomena that make interesting pictures, such as the crescent images of the partially eclipsed Sun produced by tree foliage. The narrow gaps between leaves act as "pinhole cameras" and each projects its own tiny (and inverted) image of the crescent Sun on the ground. This pinhole camera effect becomes more pronounced as the eclipse progresses.

You can make your own pinhole camera to project the crescent Sun with pinholes punched in cardboard, or by using your hands, or with a wide-brimmed straw hat. The profusion of crescents on the ground, on a building, or on a person's face makes a nice photographic memento. Almost any kind of camera will work. Just be sure to disengage the automatic flash.

Landscape Eclipse Photography

If you have an extra camera and tripod, take a few wide-angle shots of the sky, horizon, and landscape before, during, and after totality. This sequence might show the Moon's dark, fast-moving shadow approaching from the west or racing away over the horizon to the east. Place yourself or some other interesting subjects in the foreground to give the photos some scale. Full-frame fisheye lenses are especially useful and can produce dramatic images. Use your camera's light meter and bracket exposures if possible. If you know the Sun's altitude during totality, you can use the table accompanying this chapter to choose a lens focal length that will include the Sun in the

Field of View and Size of the Sun for Various Focal Lengths

Focal Length	Field of View (35 mm format)	Size of Sun on film
20 mm	69° x 103°	0.2 mm
28 mm	49° x 74°	0.3 mm
35 mm	39° x 59°	0.3 mm
50 mm	28° x 41°	0.5 mm
105 mm	13.1° x 19.6°	1.0 mm
200 mm	6.9° x 10.3°	1.8 mm
400 mm	3.4° x 5.2°	3.6 mm
500 mm	2.8° x 4.1°	4.5 mm
1000 mm	1.4° x 2.1°	9.1 mm
1500 mm	0.9° x 1.4°	13.6 mm
2000 mm	0.7° x 1.0°	18.2 mm
2500 mm	0.6° x 0.8°	22.7 mm

Note: Image size of Sun = focal length / 110

Photographing Shadow Bands

by Laurence A. Marschall

Visual observations of shadow bands are interesting and fun, but the most useful observations from a scientific standpoint are those using photography or videotape. The blotchy patterns of the bands can be predicted from mathematical models of how sunlight travels through the Earth's atmosphere. Yet few of the many attempts to catch shadow bands on film or tape have been successful. Perhaps a dozen fuzzy photographs and one flickering film are all that are available. The bands move so swiftly and have such low contrast that it is hard to get a good picture.

To photograph the bands, a high shutter speed (1/250 second or shorter) and very fast film (ISO 400 or higher) should be used. Focus the camera on a white cardboard or plywood screen laid out on the ground (see "Catching Shadow Bands" in chapter 10), and take as many shots as possible, using a range of lens openings to ensure a proper exposure. Leave a yardstick on the screen to establish a scale.

For videotaping, many handheld camcorders have a fast shutter option (1/1000 to 1/4000 second) which should be used to stop the motion of the shadow bands. If you get a distinct record of the shadow bands and their motion, it is not just something to admire; it is a remarkable scientific resource. Shadow band researchers (there are perhaps half a dozen worldwide) will beat a path to your door.

One minute after the end of totality, Canadian astrophotographer Jack Newton captured shadow bands on the wing of the DC-3 from which he observed the eclipse of July 10, 1972 over Baker Lake, Northwest Territories. [50mm focal length, f/2.8, 1/125 second, with ISO 160 film. © 1972 Jack Newton]

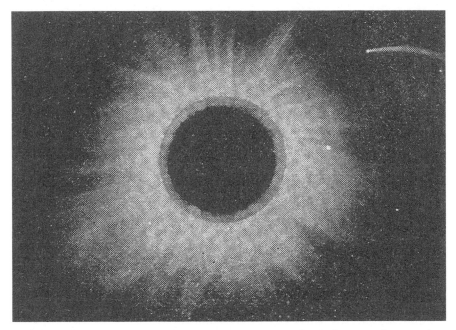

Painting made from a photograph by astronomer Arthur Schuster, who was the first to capture an unknown comet in an eclipse photograph, May 17, 1882. [Mabel Loomis Todd: *Total Eclipses of the Sun*

top of the frame. You'll also capture any bright planets near the Sun during totality. You might even discover a new comet.

Multiple Exposure Sequences

Some of the most dramatic eclipse photographs are multiple exposures that show the totally eclipsed Sun accompanied by a sequence of partial phases on either side. This kind of photograph requires a camera capable of taking multiple exposures without advancing the film. Most electronic SLRs do *not* allow this type of photography. Check with the staff of your camera store and explain what you want to do; they should be able to show you cameras that will meet your needs.

In addition to a multiple-exposure camera, you will need a sturdy tripod that will hold the camera fixed to one point in the sky without moving when you recock the shutter between exposures, and you will need a cable release to further reduce vibration.

To record all the phases of a total eclipse on one frame of film, you need to know the field of view of your camera lens. The Sun moves 15° per hour—its own diameter every two minutes. A 50 mm lens has a field of 49° along the diagonal. It should therefore take the Sun

Multiple exposure of the total eclipse of the Sun over Minneapolis on June 30, 1954. [*Star Tribune*, Minneapolis–St. Paul]

about three hours to traverse the diagonal, so try to orient your camera so that the Sun moves in this direction. From first to last contact, a total eclipse lasts about 2¼ hours. Taking an exposure every 5 to 10 minutes provides adequate separation between images on the final photograph.

If you want the image during totality to be in the center of the frame, you must calculate the times from that point in order to have all of your images equally spaced on both sides of the totally eclipsed Sun. For example, if mid-totality is at 10:32 A.M., then you would add or subtract increments of, say, 5 minutes to that figure to get the times to make your exposures. You may want to allow more of an interval just before and after totality to avoid trampling on the corona.

The altitude and azimuth of the Sun or Moon for a specific eclipse are easily obtained using astronomy computer programs available for your home computer. Other sources of information for upcoming eclipses include popular astronomy magazines, NASA's solar eclipse bulletins, and eclipse Web sites (Appendices D, E, F, and G).[19]

Eclipse Photography from Sea

For eclipse photography at sea, the pitching and rolling of the ship place certain limits on the focal length and shutter speeds that can be used. In most cases, telescopes with focal lengths of 1000 mm or longer can be ruled out because the ship would need to be virtually motionless during totality. In addition to the movement of the ship, you must also contend with vibration from the engines, wind across the deck, and hundreds of people stomping around you, so try to locate yourself to minimize these problems.

Film choice can be determined on eclipse day by viewing the Sun through the camera lens (with a solar filter) and noting the image motion caused by the ship. Some people have been successful with ISO 100 film speeds, but a safer film would be ISO 400 or faster.

When shooting pictures on a pitching or rolling ship, notice the range of motion and attempt to snap each picture at one extreme while the ship is momentarily motionless, so the image in your camera will be still.

Video Photography of Eclipses

Video cameras have provided eclipse observers with a new and easy way to preserve a dynamic celestial event. The newest cameras are smaller and lighter and come in a variety of prices and resolutions. The 8 mm format is the least expensive ($300–$500) and has the lowest resolution (250 lines). Hi-8 offers better resolution (400 lines) at about twice the price. Digital video has the highest resolution (500 lines) among consumer camcorders, but it's also the most expensive.

The new video cameras all use CCDs (charge-coupled devices), which are more sensitive to faint light but are virtually impervious to intense light that could damage older tube video cameras. CCDs can tolerate exposure to direct sunlight for brief periods of time with no harmful effects.

Zoom lenses are the norm on camcorders and have ranges like 8:1,

12:1, and even 24:1. These are all optical zoom ratios. Many cameras also have a digital zoom that magnifies the optical image without improving the resolution. Digital zooms make the image look blocky because they enlarge the pixels, so they are not recommended for quality eclipse videotaping.

A video camera with a ½-inch CCD and a 70 mm lens provides an image of the Sun about 28 mm (1.1 inches) in diameter on a 13-inch screen, large enough to produce a decent view of the diamond ring and corona, especially from ship or plane. Converter lenses are available that screw into the front of your camcorder's zoom and magnify the image by two, three, or four times. This magnification is just right for dramatic close-ups of the partial phases, the corona, and the diamond ring.

The latest camcorders include electronic image stabilization, which is especially good for videotaping eclipses from sea or air. But unless you are on a ship, your camcorder needs to be mounted on a sturdy tripod. To videotape the partial phases, you also need a solar filter. Remove the filter about 15 seconds before totality begins to capture the diamond ring effect. Some observers even remove their solar filters a full minute before second contact with no damage to the camcorder. Keep your solar filter close by so that you can replace it on your video camera about 15 seconds after totality ends.

Auto-exposure works fine for eclipses, but the inner corona and prominences will probably be overexposed unless you adjust the exposure manually. On the other hand, auto-focus often functions poorly during totality. The low light levels may cause your camcorder's lens to wander around searching for the correct focus. Turn the camcorder auto-focus off about 10 minutes before totality and focus manually. This is also a good time to put a fresh battery into your video camera.

Video photography offers amateurs a chance to contribute useful scientific data. Recording the eclipse simultaneously with an accurate time signal allows the observer to determine the precise time of totality for a given location. This information can be particularly useful at the northern and southern limits of totality, where video recordings can provide the data needed for accurate measurements of the Sun's diameter.[20]

Camcorders are also great for videotaping your observing site and the sky during totality. Just set your zoom at its widest setting and place the camcorder on a tripod. The camcorder's audio track will also capture all the conversations and excitement during the most majestic celestial event visible from planet Earth.

Some Final Words

If there is one key to successful eclipse photography, it is *preparation*. Set up and test all your equipment at home to ensure that everything works perfectly together. Design a photographic plan or schedule and stick to it. Keep things simple. Don't try to do too much. Practice for the eclipse with a full dress rehearsal. Bring extra film, batteries, cable releases, and other crucial items.

Finally, don't get so overwhelmed with taking photographs that you deprive yourself of time to actually *look* at the eclipse.

Checklist for Solar Eclipse Photography

Camera Items
Camera (two, if possible, in case one fails)
Motor drive for film advance (if available)
Lenses (16 mm to 1500 mm)
Cable release (spare advisable)
Heavy-duty tripod
Extra batteries (for camera meter and motor drive)
Film (several rolls of 36-exposure)
Sack or plastic zipper lock bags and heavy string (to fill with sand and sus-
    pend from tripod for stability)
Exposure list for partial phases and totality

Telescope Items
Telescope (consult owner's manual for photographic equipment)
    Mount, wedge, cords, eyepieces, photo adapters (T-ring, etc., to attach
        camera to telescope), counterweights, drive corrector (and spare
        fuse), lens-cleaning supplies
Solar filter (full aperture preferred)
Inexpensive solar filter for finder scope
Equatorial mount and clock drive (if available)
Magnetic compass (for polar alignment)
Bubble angle finder (to level mount and align polar axis)
Battery for telescope clock drive (if necessary)
AC converter plug, voltage corrector, and extension cord (if necessary)

Video Camera Items
Camcorder (with charger and accessories)
Tele-extender lens (2x, 3x, or 4x)
Heavy-duty tripod
Extra batteries and battery charger

General Items
Handheld solar filter (for viewing partial phases)
Penlight flashlight
Small cassette tape recorder, tapes, and batteries (to record comments dur-
    ing eclipse if not using video camera)
Pocketknife
Tools for minor repairs: screwdrivers (regular, phillips, and jeweler's), nee-
    dle-nose pliers, tweezers, small adjustable wrench, allen wrench set,
    etc.
Plastic trash bags (to protect equipment from rain or dust)
Roll of duct tape or masking tape for emergency repairs
Optional: Extra cassette recorder to play prerecorded instructions via head-
    phones
Optional: Shortwave receiver for WWV time signal or GPS (global position-
    ing system)

Photographic Settings for a Solar Eclipse

$$E = f^2 / (A \times B)$$

where: E = exposure (seconds)
f = focal ratio
A = film speed (ISO)
B = brightness value

SUN—full disk or partial eclipse
Through full-aperture solar filter
B~256
Film Speed—ISO

f/	32	64	100	200	400
2.8	1/1000	1/2000	1/4000	—	—
4	1/500	1/1000	1/2000	1/4000	—
5.6	1/250	1/500	1/1000	1/2000	1/4000
8	1/125	1/250	1/500	1/1000	1/2000
11	1/60	1/125	1/250	1/500	1/1000
16	1/30	1/60	1/125	1/250	1/500
22	1/15	1/30	1/60	1/125	1/250
32	1/8	1/15	1/30	1/60	1/125

SUN—total eclipse: inner corona (3° field)
No filter
B~32
Film Speed—ISO

f/	32	64	100	200	400
2.8	1/125	1/250	1/500	1/1000	1/2000
4	1/60	1/125	1/250	1/500	1/1000
5.6	1/30	1/60	1/125	1/250	1/500
8	1/15	1/30	1/60	1/125	1/250
11	1/8	1/15	1/30	1/60	1/125
16	1/4	1/8	1/15	1/30	1/60
22	1/2	1/4	1/8	1/15	1/30
32	1 sec	1/2	1/4	1/8	1/15

SUN—total eclipse: prominences
No filter
B~128
Film Speed—ISO

f/	32	64	100	200	400
2.8	1/500	1/1000	1/2000	1/4000	—
4	1/250	1/500	1/1000	1/2000	1/4000
5.6	1/125	1/250	1/500	1/1000	1/2000
8	1/60	1/125	1/250	1/500	1/1000
11	1/30	1/60	1/125	1/250	1/500
16	1/15	1/30	1/60	1/125	1/250
22	1/8	1/15	1/30	1/60	1/125
32	1/4	1/8	1/15	1/30	1/60

SUN—total eclipse: outer corona (10° field)
No filter
B~0.5
Film Speed—ISO

f/	32	64	100	200	400
2.8	1/2	1/4	1/8	1/15	1/30
4	1 sec	1/2	1/4	1/8	1/15
5.6	2 sec	1 sec	1/2	1/4	1/8
8	4 sec	2 sec	1 sec	1/2	1/4
11	8 sec	4 sec	2 sec	1 sec	1/2
16	15 sec	8 sec	4 sec	2 sec	1 sec
22	30 sec	15 sec	8 sec	4 sec	2 sec
32	60 sec	30 sec	15 sec	8 sec	4 sec

Note: These exposure tables are given as guidelines only. The brightness of prominences and the corona can vary considerably. You should bracket your exposures to be safe.

13

The Pedigree of an Eclipse

[T]he general phaenomenon is perhaps the most awfully grand which man can witness.
—George B. Airy (1851)

Every eclipse belongs to a family, a saros series, that begins, evolves, and ends. Each saros series has a character all its own, with distinguishing features, the ability to captivate its viewers, and a chance to participate in human history. Here is the story of one remarkable family of eclipses.

The eclipse family known as saros 136 was born on June 14, 1360, deep in the southern hemisphere, over Antarctica and the southern Indian Ocean. The Moon was at a descending node—crossing the Sun's path on its way south. The Sun just happened to be near that node at the time so that, as viewed from Earth, the Moon grazed the southwestern edge of the Sun. It was a very slight partial eclipse, unnoticeable to the eye without filters, and there was no one there to see.

It was an inauspicious birth, but the firstborn of every saros family is always a slight partial eclipse that brushes the Earth at one of the poles and gives no visual evidence of the splendor that will come as the family matures.

Celestial Clockwork

Time passed. The years rolled by. Forty-one other solar eclipses touched the Earth, but they came from other saros families. Then, after 6,585 days, the Moon had completed 223 lunations and the Sun had passed by the descending node of the Moon 19 times. The two cycles nearly matched, forcing the Sun and Moon to meet under almost the same circumstances that had prevailed 18 years 11⅓ days earlier. Another eclipse was inevitable.

But the new eclipse was not identical to the first. The Moon's node was not exactly where it had been 18 years earlier. It had backed around the Sun's orbit and was now 0.477 degrees east of its previous position, so this time the Moon cut off a slightly larger portion of the Sun's light. It was still a very slight (0.199) partial eclipse, noticeable only with optical filters in the south Atlantic Ocean and Antarctica, and again no one was there to see.

Each 18 years and a few days brought a new solar eclipse of saros family 136, each a little farther north on the whole; each a partial eclipse, but each time covering a little more of the Sun. During 126 years there were eight partial eclipses, until on August 29, 1486, the Moon's diameter covered 98.6 percent of the Sun's, leaving at maximum eclipse only a thin crescent of the Sun visible from near the south pole.

The next eclipse in the cycle, September 8, 1504, was different from those that preceded it. The Moon passed directly across the disk of the Sun as seen from near the Antarctic coast. It would have been a total eclipse if the angular size of the Moon's disk had been large enough, but the Moon's elliptical orbit had carried it farther from the Earth than average, so it appeared smaller—too small to cover the Sun completely. At maximum eclipse, the Sun's surface still shone around the circumference of the Moon to form a bright ring of light—an annular eclipse—lasting at most 31 seconds.

Throughout the sixteenth century, each of the six eclipses of saros 136 was an annular eclipse, gradually migrating northward. And with each eclipse, the Moon was a little closer to Earth, so that the Moon's apparent size grew larger and the duration of the annulus got *shorter*. On November 12, 1594, the annular eclipse lasted 4 seconds at its midpoint. The disk of the Moon was almost large enough to completely hide the Sun.

At the next eclipse, it did—for just *1 second*. On November 22, 1612, in the southeastern Pacific Ocean and Antarctica, birds, fish, and whales saw an eclipse that was annular all along the central path until the eclipse reached its midpoint, where the surface of the Earth was closest to the Moon and hence the Moon appeared largest. At that location, the Moon's disk for just an instant completely covered the Sun and the eclipse (technically, at least) became total. The dark

An occasional solar eclipse may start off annular but become total as the roundness of the Earth reaches up to intercept the shadow. The eclipse then returns to annular as the curvature of the Earth causes its surface to fall away from the shadow. These hybrids are called annular-total eclipses.

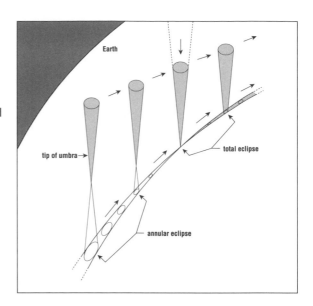

shadow of the Moon caused by saros 136 for the first time actually touched the Earth. Then, as modestly as a young boy and girl touching hands for the first time, the shadow and the Earth pulled apart as the Earth's surface curved away beneath the extended shadow. Just the slightly greater distance from the Moon that the Earth's curvature provided was enough to make the Moon once again appear too small to completely cover the Sun. The Sun peered out from around the entire circumference of the Moon, and the eclipse was annular once more.

The eclipse of 1612 and all five eclipses of saros 136 in the seventeenth century were hybrids of this kind: beginning as annular, becoming total, then returning to annular again. And with each eclipse, the duration of totality was a little longer, until on January 5, 1685, the total phase lasted 35 seconds and, at least momentarily, totality was seen for thousands of miles along the eclipse route. Only at the western and eastern extremes of the eclipse path, near sunrise and sunset, where the distance to the Moon from the surface of the Earth was slightly greater, was the eclipse annular.

The Big Leagues

On the twentieth eclipse of saros 136, the Earth was near its closest approach to the Sun, so the Sun's angular size was greatest and it was hardest to eclipse. But the Moon in its orbit was close enough to

The Extinction of Total Solar Eclipses

In 1695, Edmond Halley discovered that eclipses recorded in ancient history did not match calculations for the times or places of those eclipses. Starting with records of eclipses in his day and the observed motion of the Moon and Sun, he used Isaac Newton's new theory of universal gravitation (1687) to calculate when and where ancient eclipses should have occurred and compared them with eclipses actually observed more than 2,000 years earlier. They did not match. Halley had great confidence in the theory of gravitation and resisted the temptation to conclude that the force of gravity changed as time passed. Instead, he proposed that the length of a day on Earth must have increased by a small amount.

If the Earth's rotation had slowed down slightly, the Moon must have gained the angular momentum to conserve the total angular momentum of the Earth-Moon system. This boost in angular momentum for the Moon would have caused it to spiral slowly outward from the Earth to a more distant orbit where it travels more slowly. If, 2,000 years earlier, the Earth had been spinning a little faster and the Moon had been a little closer and orbiting a little faster, then eclipse theory and observation would match. Scientists soon realized that Halley was right.

But what would cause the Earth's spin to slow? Tides. The gravitational attraction of the Moon is the principal cause of the ocean tides on Earth. As the shallow continental shelves (primarily in the Bering Sea) collide with high tides, the Earth's rotation is retarded. The slower spin of the Earth causes the Moon to edge farther from our planet.

From 1969 to 1972, the Apollo astronauts left a series of laser reflectors on the Moon's surface. Since then, scientists on Earth have been bouncing powerful lasers off these reflectors. By timing the round trip of each laser pulse, the Moon's distance can be measured to an accuracy of several inches. The Moon is receding from the Earth at the rate of about 1.5 inches (3.8 centimeters) a year.

As the Moon recedes from Earth, its apparent disk becomes smaller. Total eclipses become rarer, annular eclipses more frequent. Total eclipses are moving toward extinction. When the Moon's mean distance from the Earth has increased by 14,550 miles (23,410 kilometers), the Moon's apparent disk will be too small to cover the entire Sun, even when the Moon's elliptical orbit carries it closest to Earth. Total eclipses will no longer be possible.

How long will that take? With the Moon receding at 1½ inches a year, the last total solar eclipse visible from the surface of the Earth will take place 620 *million* years from now. There is still time to catch one of these majestic events.

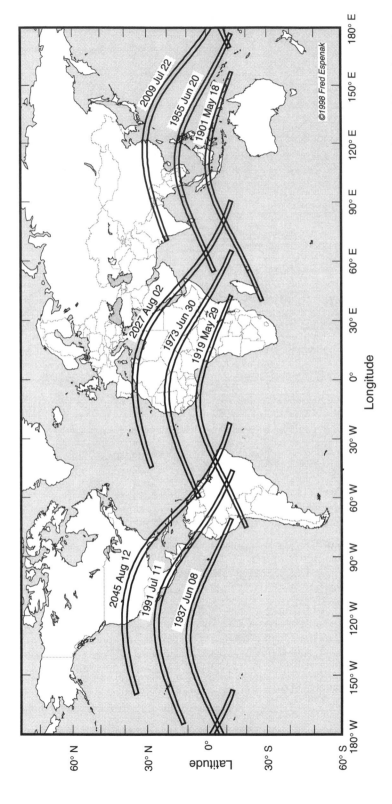

Longitude

Latitude

Paths of totality for six past and three future eclipses of saros 136. Successive eclipses shift westward and northward. For odd-numbered saroses, eclipses shift westward and *southward*. [Map and eclipse calculations by Fred Espenak]

2009 Jul 22
1955 Jun 20
1901 May 18
2027 Aug 02
1973 Jun 30
1919 May 29
2045 Aug 12
1991 Jul 11
1937 Jun 08

©1998 Fred Espenak

Earth so that its apparent size was larger than average. Thus, finally, on January 17, 1703, the eclipse was total all along the central path, with totality lasting a maximum of 50 seconds.

Each subsequent eclipse of saros 136 was total, and the duration of totality was rapidly increasing. By 1793 the eclipse lasted a maximum of 2 minutes 52 seconds over the Indian Ocean and provided a shorter show for the aborigines in Australia and the newly arrived British settlers. The eclipse path was shifting northward. As the American Civil War ended, saros 136 crossed the 5-minute plateau into the realm of rare and great eclipses as it stretched from South America to southern Africa. The totality of April 25, 1865 lasted 5 minutes 23 seconds.

Now almost everything was conspiring to make saros 136 one of the greatest in history. The eclipses were occurring later and later in the spring, when the Sun was farther from Earth and hence smaller in apparent size, enabling total eclipses to last longer. At each eclipse, the perigee of the Moon (its position closest to Earth) was close to the eclipse point and getting closer, so that the Moon was near a maximum apparent size, allowing total eclipses to last longer. Every eclipse of saros 136 in the twentieth century would last longer than 6 minutes. And the twentieth century was prepared to make good use of this rare gift.

The eclipse of May 18, 1901, brought astronomers from all over the world to Indonesia to continue their analysis of the solar atmosphere, still best seen inside the shadow of the Moon. They had at most 6 minutes 29 seconds within which to work.

A total eclipse from saros 136: July 11, 1991. Totality lasted almost 7 minutes in Baja, Mexico. [C-90 Maksutov, 1000 mm focal length, f/11, 1 second, with ISO 64 film. © 1991 Fred Espenak]

The next eclipse in the cycle lasted even longer—6 minutes 51 seconds at its peak, a totality long enough to make it memorable in and of itself. But that was not why eclipse number 32 in saros 136 became the most famous in history. Astronomers used that eclipse to measure star positions around the darkened Sun and concluded that starlight passing by the Sun had been bent, just as Einstein had predicted in his recently completed general theory of relativity. Thus the total eclipse of May 29, 1919, marked a major turning point in the history of science.

Saros 136, however, had its own dramatic schedule to fulfill, independent of the brilliance or depravity of man. It returned on June 8, 1937, as the clouds of war darkened over Europe and the Far East, with a 7-minute-4-second eclipse—the first to last over 7 minutes since 1098. Unfortunately, this amazing sight passed almost entirely across the expanse of the Pacific Ocean, avoiding land, where scientists could deploy their instruments.

Saros 136 reached its climax on June 20, 1955, with an eclipse that, at maximum, lasted 7 minutes 8 seconds, the longest in the Saros family since June 20, 1080 (7 minutes 18 seconds), and the longest until June 25, 2150 (7 minutes 14 seconds). The shadow visited Sri Lanka, Thailand, Indochina, and the Philippines. Never again would saros 136 equal that duration of totality. Yet few saroses even come close.

June 30, 1973, brought an eclipse, visible across Africa, that lasted as long as 7 minutes 4 seconds. The last offering of saros 136 in the twentieth century happened on July 11, 1991, bringing up to 6 minutes 54 seconds of totality and tracing a path from Hawaii through Mexico, Central America, Colombia, and Brazil.

The six longest eclipses of the twentieth century were in 1901, 1919, 1937, 1955, 1973, and 1991—every one of them a member of the same eclipse family. Saros 136 is one of the greatest eclipse generation cycles in recorded history.

Closing Out a Career

Now, gradually, the glory is beginning to fade. Saros 136 is declining. The Moon is receding from perigee, making its disk appear smaller; it will not cover the Sun as well. Eclipses are occurring ever later in the year, when the Earth is actually closer to the Sun so that its disk appears larger and harder to eclipse. Eclipses will become steadily shorter, but they will still be total and will still be comparatively long for quite some time. The glory will fade slowly.

Saros 136 does not end in 1991. It returns on July 22, 2009, with a 6-minute-39-second display and a central path across India, China,

The eclipse of July 11, 1991, a member of saros 136, gave astronomers a rare opportunity because its path of totality crossed Hawaii and the great observatory complex atop Mauna Kea. [Nikon F4S, 58 mm focal length, f/1.2, 1/8 second, with ISO 100 film. © 1991 Serge Brunier]

and the western Pacific. It will be back on August 2, 2027, racing over North Africa, Saudi Arabia, and the Indian Ocean, bringing darkness for up to 6 minutes 23 seconds. It will pay its first visit to the continental United States on August 12, 2045, streaking from northern California through Florida, the longest totality for the continental United States in the calculated history of eclipses. Over the Caribbean, it will last for 6 minutes 6 seconds. But that will be the last eclipse of saros 136 to surpass the 6-minute mark.

The Eclipse Family of Saros 136

Number	Date	Type	Duration*
1	1360 June 14	Partial	[0.050]
2	1378 June 25	Partial	[0.199]
3	1396 July 5	Partial	[0.346]
4	1414 July 17	Partial	[0.489]
5	1432 July 27	Partial	[0.626]
6	1450 August 7	Partial	[0.757]
7	1468 August 18	Partial	[0.876]
8	1486 August 29	Partial	[0.986]
9	1504 September 8	Annular	0:31
10	1522 September 19	Annular	0:23
11	1540 September 30	Annular	0:17
12	1558 October 11	Annular	0:12
13	1576 October 21	Annular	0:08
14	1594 November 12†	Annular	0:04
15	1612 November 22	Annular/total	totality: 0:01
16	1630 December 4	Annular/total	totality: 0:07
17	1648 December 14	Annular/total	totality: 0:15
18	1666 December 25	Annular/total	totality: 0:24
19	1685 January 5	Annular/total	totality: 0:35
20	1703 January 17	Total	0:50
21	1721 January 27	Total	1:07
22	1739 February 8	Total	1:28
23	1757 February 18	Total	1:52
24	1775 March 1	Total	2:20
25	1793 March 12	Total	2:52
26	1811 March 24	Total	3:27
27	1829 April 3	Total	4:05
28	1847 April 15	Total	4:44
29	1865 April 25	Total	5:23
30	1883 May 6	Total	5:58
31	1901 May 18	Total	6:29
32	1919 May 29	Total	6:51
33	1937 June 8	Total	7:04
34	1955 June 20	Total	7:08
35	1973 June 30	Total	7:04
36	1991 July 11	Total	6:53
37	2009 July 22	Total	6:39
38	2027 August 2	Total	6:23
39	2045 August 12	Total	6:06

*Duration of totality or annularity is given in minutes:seconds. If eclipse is partial, its magnitude (fraction of Sun's diameter covered by Moon) appears in brackets.

†Henceforth on the Gregorian calendar, which replaced the Julian calendar in 1582 and dropped 10 days from that year to keep March at the beginning of spring.

Number	Date	Type	Duration*
40	2063 August 24	Total	5:49
41	2081 September 3	Total	5:33
42	2099 September 14	Total	5:18
43	2117 September 26	Total	5:04
44	2135 October 7	Total	4:50
45	2153 October 17	Total	4:36
46	2171 October 29	Total	4:23
47	2189 November 8	Total	4:10
48	2207 November 20	Total	3:56
49	2225 December 1	Total	3:43
50	2243 December 12	Total	3:30
51	2261 December 22	Total	3:17
52	2280 January 3	Total	3:04
53	2298 January 13	Total	2:52
54	2316 January 25	Total	2:42
55	2334 February 5	Total	2:32
56	2352 February 16	Total	2:24
57	2370 February 27	Total	2:17
58	2388 March 9	Total	2:10
59	2406 March 20	Total	2:03
60	2424 March 31	Total	1:55
61	2442 April 11	Total	1:46
62	2460 April 21	Total	1:34
63	2478 May 3	Total	1:21
64	2496 May 13	Total	1:02
65	2514 May 25	Partial	[0.952]
66	2532 June 5	Partial	[0.824]
67	2550 June 16	Partial	[0.685]
68	2568 June 26	Partial	[0.544]
69	2586 July 7	Partial	[0.397]
70	2604 July 19	Partial	[0.252]
71	2622 July 30	Partial	[0.105]

Breakdown of solar eclipses in saros 136:

Total	45
Annular	6
Annular/total	5
Partial	15

Solar eclipses in saros 136:   71

Span of saros 136:   1,262 years

Note: Saroses with even numbers occur at the descending node; they start near the south pole and progress northward. Saroses with odd numbers occur at the ascending node; they start near the north pole and progress southward.

Saros Series Statistics

	Range	Average
Number of solar eclipses in a saros series	70–85	73
Time span for a series (approximate)	1,244–1,514 years	1,315 years

At any time, about 42 saros series are running simultaneously.

Three saros cycles (54 years) later, on September 14, 2099, it will return to North America, slicing across southwestern Canada and plunging southeastward across the United States to the mouth of the Chesapeake Bay and off into the Atlantic. At maximum, east of the Windward Islands, the eclipse will last 5 minutes 18 seconds.

Saros 136 is aging, but its eclipses are still total and more than long enough to lure people by the thousands into its shadows. By May 13, 2496, however, a total eclipse of saros 136 will have dwindled to 1 minute 2 seconds for hardy travelers in the Arctic. It is the last of 45 total eclipses in this remarkable saros family.

At the next visit of this saros, May 25, 2514, the dark shadow of the Moon will miss the Earth, passing above the north polar region. The remaining six eclipses in the sequence will all be partial as well, steadily declining in the Moon's coverage of the face of the Sun. On July 30, 2622, there will be one final partial eclipse, virtually unnoticeable near the north pole. Saros 136 will have died.

Saros number 136 will have performed on Earth for 1,252 years and created 71 solar eclipses. Not an exceptionally long career, but what a record! Of its 71 eclipses, 15 will have been partial, 6 annular, 5 a combination annular and total, and *45 total*. The typical saros offers an average of only 19 or 20 total eclipses.

Saros 136 brought the people of the twentieth century three eclipses with totality exceeding 7 minutes and all six of the longest eclipses in that century. Its three eclipses in the first half of the twenty-first century will all exceed 6 minutes, the longest eclipses to be seen in that span of time. If this saros were an athlete, its shadow-black jersey with its corona-white number 136 would be retired to hang in glory in the Eclipse Hall of Fame.

14

The Eclipse of August 11, 1999

[A] total eclipse of the Sun . . . is the most sublime and awe-inspiring sight that nature affords.

—Isabel Martin Lewis (1924)

The Moon's shadow has darkened space for 4½ billion years, since the Sun formed and began to shine and the planets and their moons formed and could not shine. From the newly formed Sun, light streamed outward in all directions. Here and there it illuminated a body of rock or gas or ice. The Sun's dominant light identified that world as one of its own children.

A little over eight minutes of light-travel time outbound from the Sun, a portion of the light encountered two small dark bodies and bathed the sunward half of each in brightness. The surrounding flood of unimpeded light sped on, leaving a long cone of darkness—a shadow—behind the planet and its one large moon.

The two worlds lay the same average distance from the Sun. Not long before—measured on a cosmic timescale—Earth and Moon had been a single body, but they were separate and utterly different now.

A Cosmic Birth

The Sun, planets, moons, comets, and asteroids had all begun within a cloud of gas and dust, a cloud so large that there was material enough to make dozens or hundreds of solar systems. Where the density was great enough, fragments of the cloud began to condense by gravity. At the heart of each fragment, a star (or two or more) was coalescing. Near those stars-to-be, other bodies, too low in mass ever to reach starhood, also began to form. These small bodies were planetesimals, the beginnings of the planets. At first they grew by gentle collisions and adhesions, gathering up a grain of ice, a fleck of dust along their paths around the Sun until icy or rocky planetesimals had taken shape. And still they gathered dust and small debris until they were so massive that gravity became their prime means of growth, gathering to them still more materials—and other planetesimals. The number of small bodies declined. The size of a few large bodies increased. The planets had been formed by convergence.

Convergence brought near disaster as well. Another planet-size body (the size of Mars today) wandered across the path of a body that would one day be called the Earth. No living thing witnessed the collision. No living thing could have survived the collision that obliterated the smaller world and nearly shattered the Earth. The Earth recoiled from the impact by spewing molten fragments of its crust and mantle outward. Some escaped from Earth; others rained down from the skies, pelting the surface in a rock storm of unimaginable proportions. But many of the fragments, caught in the Earth's gravity, stayed aloft, orbiting the Earth as the Earth orbited the Sun. Quickly, in only 10,000 years or so, the fragments joined together by collisions and accretion to form a new world circling the first. That new world was the Moon.[1]

From convergence had come divergence. The two worlds, sprung from one, continued to diverge. Both were the same distance from the Sun, but the Earth was 81 times more massive than the Moon. That mass allowed the Earth to hold an atmosphere by gravity, while the Moon could not.

The eons passed. Life arose on Earth and covered the planet. Plants and animals responded to the tides raised by the Moon. The lunar tides slowed the Earth and caused the Moon to spiral slowly outward, diverging ever farther from the Earth in distance and ever further from the Earth in environment as well.

The lifeless Moon withdrew until today its shadow can just barely reach the Earth. As shadows in the universe go, this one is of no great size: a cone of darkness extending at most only 236,000 miles (379,900 kilometers) in length before dwindling to a point. It is long enough to touch the Earth only occasionally and very briefly and with a single narrow stroke.

The black insubstantial cone reaches out, but for most of the time there is nothing to touch. The shadow sweeps on through space unseen, unnoticed.

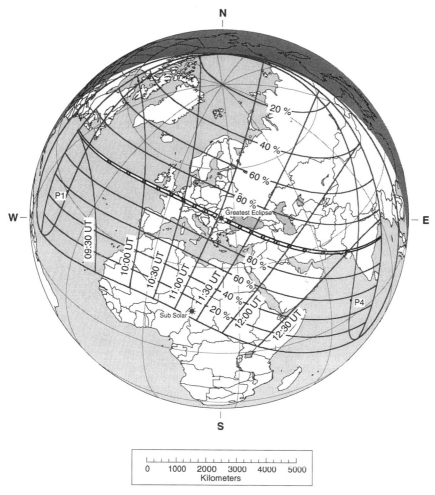

The total solar eclipse of August 11, 1999: path of totality and contours of partiality, showing percentages. [Maps in this chapter by Fred Espenak]

Contact

Yet now ahead lies the Earth. It is a special day: August 11, 1999. Suddenly the Moon's shadow becomes visible as a point of darkness collides with a world of rock and water and air. The shadow swoops in from the heavens, silently darkening the sky as it alights upon the Earth, on this occasion upon the western Atlantic Ocean, and begins its ceaseless rush across the Earth's surface.

The shadow strikes the Earth at sunrise, 300 miles (500 kilometers)

The path of the Moon's shadow during the total solar eclipse of July 11, 1991, captured in this composite of eight NASA GOES-7 weather satellite images. [Courtesy William Emery and Timothy D. Kelley, University of Colorado]

east of Boston. There, on the eastern horizon just above the waves, the Sun rises in total eclipse. Totality lasts 47 seconds.

Seen from space, the shadow sweeps eastward across the Atlantic like a giant black felt-tip marker whose ink vanishes as it moves. The shadow makes landfall for the first time on this journey 39 minutes and 2,900 miles (4,600 kilometers) after it touched down. It slides ashore on the Isles of Scilly and then the southwesternmost tip of the British mainland at Land's End. For 2 minutes, the Sun plays pirate at Penzance, a black patch covering its dazzling eye. The eclipse skims over Plymouth and then swims the English Channel to Cherbourg, France, about 90 miles (145 kilometers) in just over 2 minutes, where it invades Normandy. Le Havre and Rouen quickly fall under its shadow.

From all over the world, people have gathered along a track that runs diagonally through Europe and southwestern Asia on which a different kind of *Orient Express* will dash this day along a more northerly route from England to Turkey . . . and keep on going. This *Orient Express*, highballing in excess of 1,500 miles per hour (2,400 kilometers per hour), runs this day only, makes no stops, picks up no passengers. But people from all corners of the globe have come to watch it pass. They have come to stand in the shadow of the Moon to

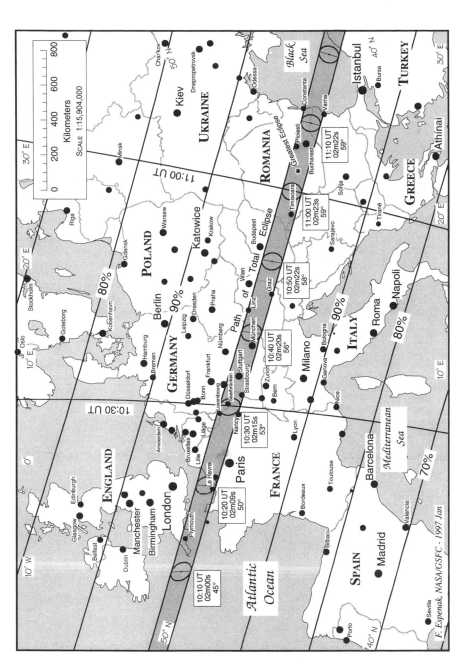

Path of totality through Europe, August 11, 1999.

1999 Total Solar Eclipse Contact Times in Europe

Location Name	First (h) (m)	Contacts Second (h) (m)	Contacts Third (h) (m)	Fourth (h) (m)	Sun Altitude (degrees)	Duration of Totality
Austria						
Graz	09:22	10:45	10:46	12:09	58	01m 12s
Linz	09:21	10:43	10:43	12:06	57	00m 30s
Salzburg	09:18	10:40	10:42	12:04	57	02m 02s
Bulgaria						
Dobric	09:45	11:10	11:12	12:32	59	02m 05s
England						
Falmouth	08:57	10:11	10:13	11:32	46	02m 02s
Plymouth	08:58	10:13	10:15	11:34	46	01m 39s
Torquay	08:59	10:14	10:15	11:35	47	01m 07s
France						
Amiens	09:05	10:22	10:24	11:45	51	01m 51s
Hagondange	09:09	10:28	10:30	11:51	53	02m 14s
Le Havre	09:02	10:19	10:20	11:41	50	01m 31s
Metz	09:09	10:28	10:30	11:52	53	02m 13s
Reims	09:07	10:25	10:27	11:48	52	01m 59s
Rouen	09:03	10:20	10:22	11:43	50	01m 36s
Strasbourg	09:11	10:31	10:32	11:55	54	01m 24s
Thionville	09:09	10:28	10:30	11:51	53	02m 07s
Germany						
Augsburg	09:15	10:36	10:38	12:00	56	02m 17s
Göppingen	09:14	10:34	10:36	11:58	55	02m 17s
Heilbronn	09:13	10:33	10:35	11:57	54	01m 29s
Ingolstadt	09:16	10:37	10:39	12:01	56	01m 24s
Karlsruhe	09:12	10:32	10:34	11:55	54	02m 08s
Munich	09:16	10:37	10:39	12:01	56	02m 08s
Neunkirchen/Saar	09:11	10:30	10:32	11:53	53	01m 53s
Pforzheim	09:13	10:32	10:34	11:56	54	02m 15s
Reutlingen	09:13	10:33	10:35	11:57	55	01m 56s
Saarbrücken	09:10	10:29	10:31	11:53	53	02m 09s
Saarlouis	09:10	10:29	10:31	11:52	53	02m 00s
Stuttgart	09:13	10:33	10:35	11:57	55	02m 17s
Ulm	09:14	10:34	10:37	11:59	55	02m 05s
Zweibrücken	09:11	10:30	10:32	11:53	54	02m 02s
Hungary						
Kecskemét	09:29	10:53	10:54	12:16	58	01m 25s
Szeged	09:30	10:53	10:56	12:17	59	02m 21s
Székesfehérvár	09:27	10:50	10:52	12:13	58	01m 37s
Luxembourg						
Luxembourg	09:10	10:28	10:30	11:51	53	01m 20s
Romania						
Arad	09:32	10:56	10:58	12:19	59	02m 14s
Bucharest	09:41	11:06	11:08	12:29	59	02m 22s
Pitesti	09:39	11:03	11:05	12:26	59	02m 23s
Rîmnicu-Vîlcea	09:38	11:02	11:04	12:25	59	02m 22s
Timisoara	09:32	10:56	10:58	12:20	59	02m 02s
Yugoslavia						
Subotica	09:29	10:53	10:55	12:17	59	01m 42s

Note: All times are Greenwich Mean Time (GMT).

1999 Total Solar Eclipse Contact Times in the Middle East and Asia

Location Name	Contacts				Sun Altitude (degrees)	Duration of Totality
	First (h) (m)	Second (h) (m)	Third (h) (m)	Fourth (h) (m)		
India						
Akola	11:31	12:33	12:34	—	11	00m 55s
Anand	11:26	12:31	12:32	13:29	15	00m 49s
Bhusāwal	11:30	12:33	12:34	—	12	00m 47s
Burhānpur	11:30	12:33	12:34	—	12	00m 46s
Chandrapur	11:33	12:34	12:35	—	8	00m 52s
Jālgaon	11:30	12:33	12:34	13:30	12	00m 26s
Nadiād	11:26	12:31	12:31	13:29	16	00m 25s
Surendranagar	11:25	12:30	12:31	13:29	17	01m 02s
Vadodara	11:27	12:31	12:32	13:30	15	01m 02s
Yavatmāl	11:32	12:34	12:35	—	10	00m 55s
Iran						
Aligūdarz	10:40	11:59	12:01	13:11	43	01m 51s
Bam	11:01	12:15	12:16	13:21	33	01m 26s
Borūjerd	10:38	11:57	11:59	13:10	44	01m 53s
Esfahān	10:45	12:03	12:04	13:13	41	01m 33s
Khomeynīshahr	10:45	12:02	12:04	13:13	41	01m 39s
Nahāvand	10:37	11:56	11:58	13:09	44	01m 54s
Najafābād	10:44	12:02	12:04	13:13	41	01m 47s
Qomsheh	10:46	12:04	12:06	13:14	40	01m 34s
Rafsanjān	10:56	12:11	12:12	13:18	35	01m 20s
Sanandaj	10:32	11:53	11:54	13:06	46	01m 15s
Iraq						
Al-Mawṣil (Mosul)	10:23	11:47	11:47	13:02	50	00m 30s
As-Sulaymāniyah	10:29	11:50	11:52	13:05	48	01m 56s
Irbil	10:25	11:47	11:49	13:03	49	01m 50s
Pakistan						
Karachi	11:18	12:26	12:27	13:27	22	01m 13s
Turkey						
Amasya	10:03	11:28	11:30	12:47	56	02m 10s
Batman	10:17	11:40	11:42	12:57	52	02m 07s
Cizre	10:20	11:43	11:45	12:59	51	02m 06s
Çorum	10:02	11:27	11:28	12:46	56	01m 51s
Diyarbakir	10:16	11:39	11:41	12:56	53	01m 20s
Elaziğ	10:13	11:36	11:38	12:54	53	02m 04s
Kastamonu	09:59	11:23	11:25	12:44	57	02m 17s
Siirt	10:19	11:42	11:43	12:58	51	01m 29s
Silvan	10:17	11:40	11:42	12:57	52	02m 04s
Sivas	10:07	11:31	11:33	12:50	55	02m 07s
Tokat	10:05	11:29	11:32	12:48	55	02m 09s
Turhal	10:04	11:28	11:31	12:48	56	02m 15s

Note: All times are Greenwich Mean Time (GMT).

A Young Saros at Work:
The Total Eclipse of August 11, 1999

A total eclipse that peaks at middle latitudes belongs either to a young saros whose central eclipses are just beginning or to an old saros whose flashiest eclipses are dwindling to an end. An eclipse whose greatest occultation occurs near the equator indicates a saros at its prime.

Saros 145 created the eclipse of August 11, 1999. This family of eclipses is just 360 years into a dynasty that will span 1,370.3 years. Since its first

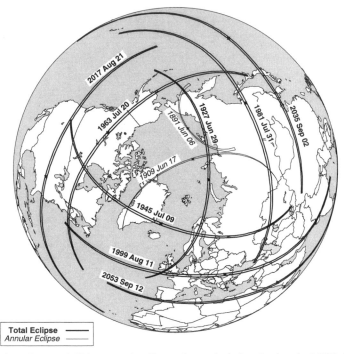

Total Eclipse ——————
Annular Eclipse ——————

The paths of saros 145's central eclipses from their beginning in 1891 through 2053. Each eclipse is westward and southward of the one before it. Saros 145 created the 1999 eclipse across Europe and southwestern Asia. Its next return will bring the 2017 eclipse for the United States. [Map and eclipse calculations by Fred Espenak]

watch the blinding bright face of the Sun vanish in broad daylight and reveal its feathery white outer atmosphere (the corona) and the scarlet filaments (prominences) that hover at the base of the corona, just above the rim of the Moon.

Now and only now, in a total eclipse of the Sun, are these features revealed to the creatures of Earth without need of highly specialized equipment. Time after time, experienced professional astronomers,

eclipse in 1639 near the north pole, it has been migrating southward. Saros 145 only recently entered adulthood with the commencement of its central eclipses: an annular eclipse in 1891, an annular-total hybrid in 1909, and its first total eclipse in 1927. The August 11, 1999, eclipse is number 21 of 77 eclipses in the series and just the fifth of 41 total eclipses saros 145 will produce. It is gathering strength with each return as it edges toward the equator. Saros 145 has progressed so that the longest totality for the 1999 eclipse takes place at latitude 45° north.

On June 25, 2522, saros 145 will offer its longest totality, 7 minutes 12 seconds in duration. As it pushes deeper into the southern hemisphere, its total eclipses will end in 2648 and its last eclipse, a partial, will take place in 3009.

The spectacle saros 145 brings in 1999 will enchant millions of people who live along its path of totality and millions more who will travel a few miles or thousands of miles to watch the Sun and Moon perform their celestial magic. But for saros 145, its 1999 tour of Europe and southwestern Asia is more than a crowd-pleasing performance. It is also a preview of coming attractions.

The next time saros 145 stages an eclipse on Earth is August 21, 2017, when the path of totality will cut diagonally from northwest to southeast across the United States (map on page 26), bringing the first total eclipse to America since 1991 (in Hawaii) and the first total eclipse for the continental United States since 1979. From 1979 to 2017: 38 years is an unusually long drought of total eclipses for an area so large.

As one millennium ends and a new one begins, saros 145 is coming of age.*

Summary of Saros 145

Total eclipses 41
Annular eclipses 1
Annular-total eclipses 1
Partial eclipses 34

---

*Saros data from Fred Espenak and Jay Anderson: *Total Solar Eclipse of 1999 August 11* (Greenbelt, Maryland: National Aeronautics and Space Administration Goddard Space Flight Center [NASA Reference Publication 1398], 1997), pages 12, 104.

---

veterans of eclipse pilgrimages, have forgotten their research assignments and stood frozen in awe at what they proclaim to be the single most beautiful, most unforgettable sight in all the heavens.

All along the eclipse path, people have been assembling, talking excitedly, preparing anxiously, watching as the shadow nears and the Sun is eaten away cookielike, bite by bite, by the occulting Moon. Submerged completely in the shadow of the Moon, they stand hushed before the actual sight, except perhaps for an unconscious murmur of wonder.

The eclipse passes just north of Paris and through the Champagne region, depriving the grapes of sunlight for 2¼ minutes. Some bottles of this year's champagne will be labeled "Eclipse." The shadow ignores borders, pressing on through southernmost Belgium and the capital of Luxembourg. It enters Germany near Saarbrücken 3 minutes before it leaves France at Strasbourg. Stuttgart lies precisely on the center line of the eclipse. Its two-thirds of a million residents experience a totality that has expanded to 2 minutes 17 seconds. Munich, with 1½ million people, is just south of the center line but entirely within the limits of the total eclipse.

The Earth is 5 weeks past aphelion, its greatest distance from the Sun, so the Sun's apparent disk is smaller than average and easier to cover. Meanwhile, the Moon is just 3½ days past perigee, its least distance from Earth, so its apparent disk is larger than average. It is 2.9 percent larger in angular size than the Sun. This moderate overlap creates a totality of moderate length. Because the shadow is only moderately wide, both the sky and ground around observers won't be exceptionally dark, but the smaller apparent size of the Moon covering the Sun may make it possible to see vibrant red prominences in the Sun's lower atmosphere throughout the entire period of totality. When the Moon is closer, and therefore larger in the sky and able to hide the Sun better, eclipses are longer and the sky is darker, but the lower regions of the corona and the chromosphere where the prominences dwell can then be seen only at the beginning of totality, when the eastward-moving Moon has just barely covered the eastern edge of the Sun, and at the end of totality, when the Moon is about to uncover the western edge of the Sun. The sunspot cycle, which averages 11 years, reached its most recent minimum in May 1996, so sunspots are again on the increase, meaning that the August 11, 1999, eclipse should exhibit more prominences and a corona that has the more extended and rounder features typical of sunspot maximum.

The eclipse flies onward from Germany into Austria, covers Salzburg, scales the mountainous spine of the country, and waltzes a little south of Vienna. It nips the northeastern corner of Slovenia and glides rapidly through southern Hungary across the Danube south of Budapest, clipping the northeastern corner of Yugoslavia.

The width of the shadow and the length of totality have been gradually increasing, and now, as the eclipse tours Romania and traverses the Transylvanian Alps, the eclipse reaches its maximum. A

Equipment Checklist for Eclipse Day

Checklist of your intended activities during eclipse

Camera equipment

Binoculars and/or telescope

Solar filters for eyes

Solar filters for binoculars and/or telescope

Thin sheets of cardboard (or a box) prepared to be a pinhole camera for indirect viewing of partial phases

Portable seat or ground covering

Flashlight with new batteries

Pencil and paper to record impressions or to sketch (also to take down the names and addresses of fellow observers)

Suitable clothing and hat (you will be outside for several hours)

Sunglasses (*not* for direct viewing of partial phases)

Bug repellent, sunscreen lotion, basic first-aid kit

Snacks and a canteen of water

Tape recorder into which you can quietly dictate your impressions of the eclipse or capture those of people near you or the reaction of wildlife

Tape recorder with earphones and prerecorded tape timed to cue you about what you wanted to do next (to run from about 2 minutes before totality until 2 minutes after totality)

---

few steps from the town of Rîmnicu-Vîlcea in south-central Romania, the duration of totality is longest—2 minutes 23 seconds. Here too the eclipse path is its widest, 69.8 miles (112.3 kilometers) across.

At this moment in this eclipse, the Sun stands closest to the zenith with the Moon covering its face and the cone of the lunar shadow aimed most nearly at the center of the Earth.[2] This, technically, is *greatest eclipse*, when the centers of the Sun, Moon, and Earth are most closely aligned. Here the eclipse shadow is moving the slowest it will travel on this passage—1,519 miles (2,444 kilometers) per hour.

Now, with the Earth's spherical surface sloping away from the Moon's shadow cone, the eclipse path begins to narrow and the length of totality begins to wane. But it has scarcely lost a second in duration when the center line of the eclipse races directly through Bucharest, capital of Romania and home to 2¼ million people. Once

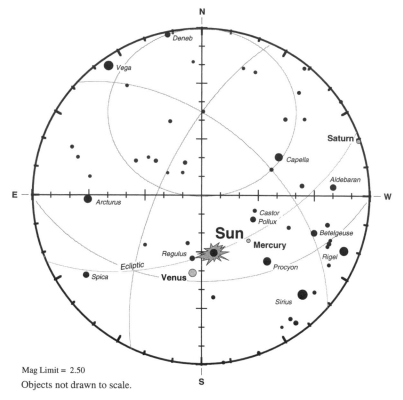

Planets and bright stars around the Sun during the eclipse of August 11, 1999. This map was plotted for the Sun highest in the sky (Romania), but the star field will look much the same from all locations along the path of totality, except that the horizon will obscure part of the western sky for viewers in Iran, Pakistan, and India. Venus (magnitude -3.5), 15° east of the Sun, will outshine all the stars. Mercury (+0.7), 18° west of the Sun, may be visible. [Map by Fred Espenak]

more the eclipse shadow crosses the Danube as it shears the north-eastern corner of Bulgaria, then sails out across the Black Sea.

Ever Onward

Through Europe, as now through Turkey and along the remainder of the eclipse path, telescopes and cameras stand like soldiers at attention, but with eyes upward. Except to adjust alignment or reload film, the people stand transfixed beside their instruments, like them staring skyward.

As the eclipse reaches Turkey, the shadow roller-coasters over the coastal mountains, plunging southeastward through eastern Turkey, rippling over the headwaters of the Tigris and Euphrates Rivers.

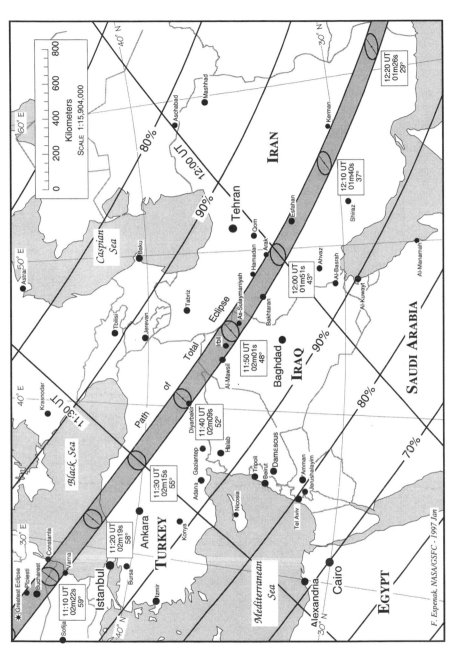

Path of totality through the Middle East, August 11, 1999.

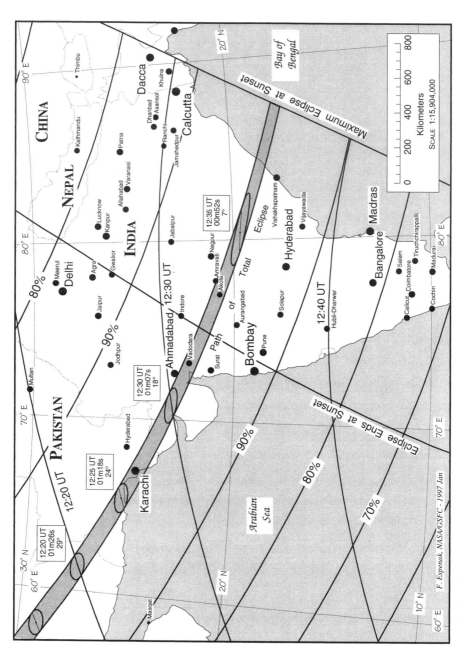

Path of totality through southwestern Asia, August 11, 1999.

The eclipse ignores the no-fly zone in northern Iraq, where it picks up the Zagros Mountain Range and tracks its entire length across the longest dimension of Iran, gradually drifting off onto the high plateau of the Salt Desert. Upon entering Iran, the total phase of the eclipse drops below 2 minutes in duration. As the shadow begins to fall off the edge of the Earth, it slides along faster and faster. Onward it rushes through southern Pakistan and directly over its largest city, Karachi, where 6 million people watch the sky grow dark for a minute and a quarter.

Then, as the Sun sinks toward a setting in the west, with scarcely more than a minute of totality left and that slipping away, the shadow, gathering speed, races across central India to leap off the surface of the Earth 200 miles (325 kilometers) offshore in the Bay of Bengal.

And it is gone.

This encounter with a shadow is over. The Moon's shadow will brush the Earth again in the eclipse seasons ahead. But there will not be an eclipse quite like this one. Each is unique.

This one grazed and graced the Earth for 3 hours 6 minutes along a path 8,490 miles (13,670 kilometers) long. The shadow has drawn its curving line across less than one-quarter of one percent of the Earth's surface, but has touched 50 million people, the most ever covered by the shadow of the Moon, the most ever to see the Sun extinguished in the midst of the day.[3]

A total eclipse of the Sun. Special are the people privileged to witness this event and feel its power. The shadow may vanish but its memory is indelible.

15

Coming Attractions, 2000–2020

Maps for Every Eclipse from 2000 through 2020

Between the years 2000 and 2020, a 21-year period, the Moon will eclipse the Sun 48 times. This interval samples at least one eclipse from every saros series currently producing eclipses. The eclipses during this period fall into the following categories:

			2000–2020	Compared to 1501–2500
Total	13	=	27.1%	27.4%
Annular	15	=	31.2%	33.3%
Annular-Total (Hybrid)	2	=	4.2%	4.3%
Partial	18	=	37.5%	35.0%

The following pages offer 48 global maps, one for each eclipse. The odd saddle-shaped zone in each map shows the region where the partial eclipse is visible. The magnitude of each eclipse (maximum fraction of the Sun's diameter covered) is shown in increments of 25 percent, 50 percent, and 75 percent. This allows you to quickly estimate the magnitude for any location within the eclipse path. For central eclipses, the path of either totality or annularity is plotted.

The Moon's penumbral shadow typically produces a zone of par-

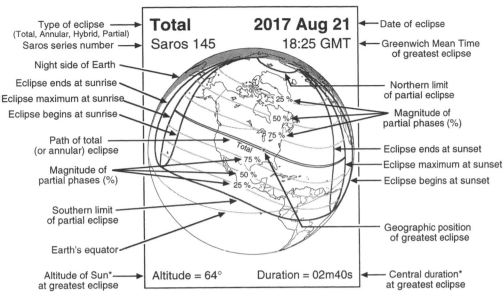

Type of eclipse (Total, Annular, Hybrid, Partial)
Saros series number

Total   **2017 Aug 21**
Saros 145   18:25 GMT

Date of eclipse
Greenwich Mean Time of greatest eclipse

Night side of Earth
Eclipse ends at sunrise
Eclipse maximum at sunrise
Eclipse begins at sunrise

Northern limit of partial eclipse

25 %
50 %
75 %

Magnitude of partial phases (%)

Path of total (or annular) eclipse

Total
75 %
50 %
25 %

Eclipse ends at sunset
Eclipse maximum at sunset
Eclipse begins at sunset

Magnitude of partial phases (%)

Southern limit of partial eclipse

Earth's equator

Geographic position of greatest eclipse

Altitude of Sun* at greatest eclipse

Altitude = 64°   Duration = 02m40s

Central duration* at greatest eclipse

* Altitude and duration are given for central eclipses.
(Eclipse magnitude is given for partial eclipses.)

tial eclipse (during both partial and central eclipses) that covers 25% to 50% of the daylight hemisphere of the Earth. In comparison, the Moon's umbral shadow (total eclipse) or antumbral shadow (annular eclipse) is much smaller: its path covers less than 1% of the Earth's surface.

Additional information on each map can be identified using the key above.

Total and Annular Eclipse Paths from 2000 through 2020

On pages 204–206 are three maps: North and South America, Europe and Africa, and Asia and Australia. These maps show in detail the paths of every central solar eclipse (total, annular, and annular-total) from 2001 through 2020.

Remember, words and pictures can never fully convey the wonder of a total eclipse of the Sun: to stand in the path of totality, in the light of the corona. You must experience one for yourself—or two, or . . .

Hope to see you there.

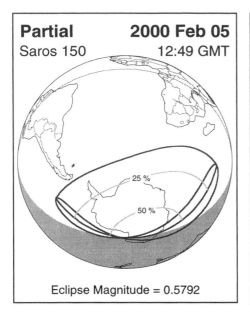

Partial   **2000 Feb 05**
Saros 150   12:49 GMT

25 %

50 %

Eclipse Magnitude = 0.5792

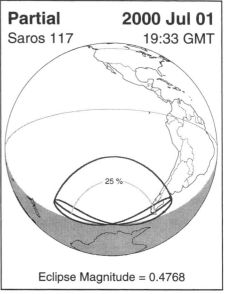

Partial   **2000 Jul 01**
Saros 117   19:33 GMT

25 %

Eclipse Magnitude = 0.4768

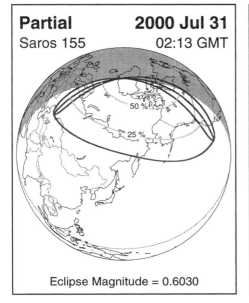

Partial   **2000 Jul 31**
Saros 155   02:13 GMT

50 %

25 %

Eclipse Magnitude = 0.6030

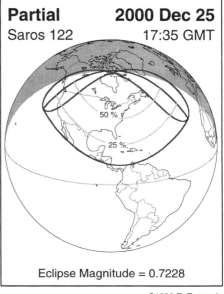

Partial   **2000 Dec 25**
Saros 122   17:35 GMT

50 %

25 %

Eclipse Magnitude = 0.7228

©1998 F. Espenak

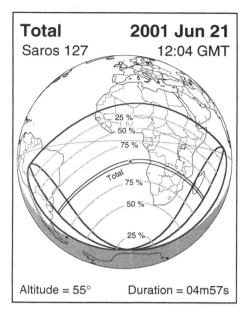

Total   2001 Jun 21
Saros 127   12:04 GMT

Altitude = 55°   Duration = 04m57s

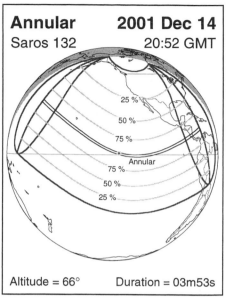

Annular   2001 Dec 14
Saros 132   20:52 GMT

Altitude = 66°   Duration = 03m53s

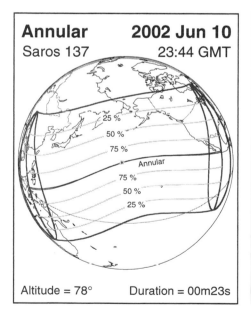

Annular   2002 Jun 10
Saros 137   23:44 GMT

Altitude = 78°   Duration = 00m23s

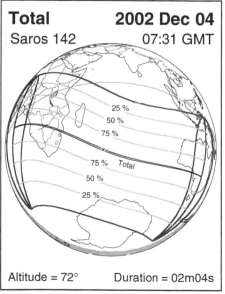

Total   2002 Dec 04
Saros 142   07:31 GMT

Altitude = 72°   Duration = 02m04s

©1998 F. Espenak

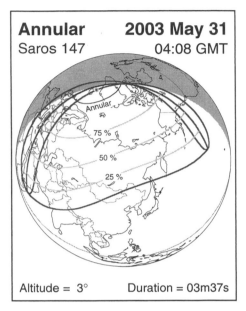

Annular   2003 May 31
Saros 147   04:08 GMT

Altitude = 3°   Duration = 03m37s

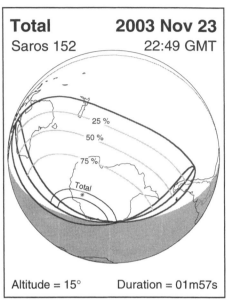

Total   2003 Nov 23
Saros 152   22:49 GMT

Altitude = 15°   Duration = 01m57s

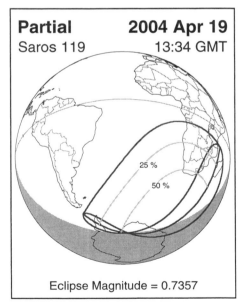

Partial   2004 Apr 19
Saros 119   13:34 GMT

Eclipse Magnitude = 0.7357

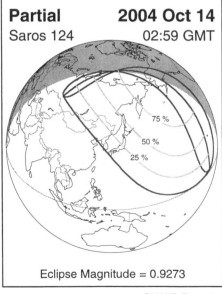

Partial   2004 Oct 14
Saros 124   02:59 GMT

Eclipse Magnitude = 0.9273

©1998 F. Espenak

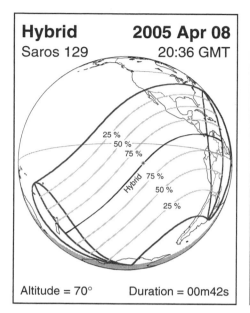

Hybrid   **2005 Apr 08**
Saros 129   20:36 GMT

25 %
50 %
75 %
Hybrid 75 %
50 %
25 %

Altitude = 70°   Duration = 00m42s

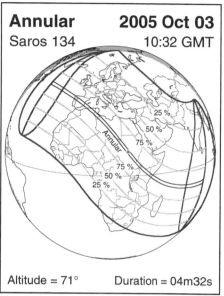

Annular   **2005 Oct 03**
Saros 134   10:32 GMT

25 %
50 %
Annular 75 %
75 %
50 %
25 %

Altitude = 71°   Duration = 04m32s

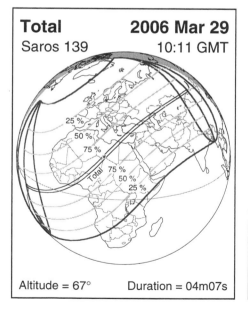

Total   **2006 Mar 29**
Saros 139   10:11 GMT

25 %
50 %
75 %
Total 75 %
50 %
25 %

Altitude = 67°   Duration = 04m07s

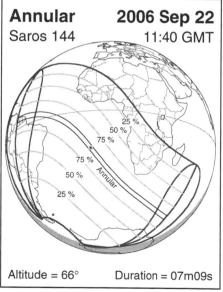

Annular   **2006 Sep 22**
Saros 144   11:40 GMT

25 %
50 %
75 %
75 %
50 % Annular
25 %

Altitude = 66°   Duration = 07m09s

©1998 F. Espenak

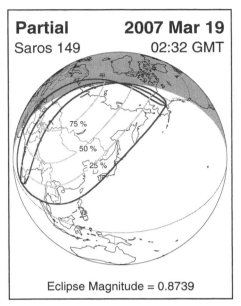

Partial   **2007 Mar 19**
Saros 149   02:32 GMT

75 %
50 %
25 %

Eclipse Magnitude = 0.8739

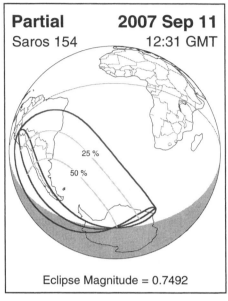

Partial   **2007 Sep 11**
Saros 154   12:31 GMT

25 %
50 %

Eclipse Magnitude = 0.7492

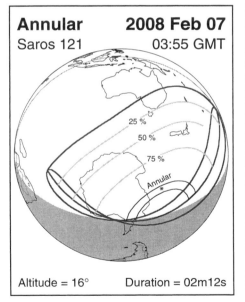

Annular   **2008 Feb 07**
Saros 121   03:55 GMT

25 %
50 %
75 %
Annular
*

Altitude = 16°   Duration = 02m12s

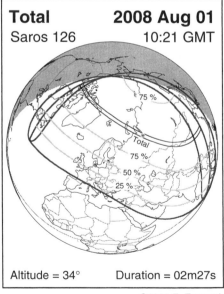

Total   **2008 Aug 01**
Saros 126   10:21 GMT

75 %
Total
75 %
50 %
25 %

Altitude = 34°   Duration = 02m27s

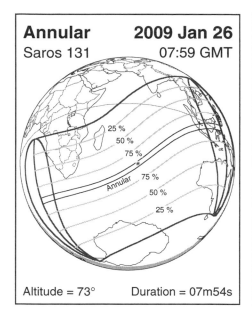

Annular   **2009 Jan 26**
Saros 131   07:59 GMT

25 %
50 %
75 %
Annular   75 %
50 %
25 %

Altitude = 73°   Duration = 07m54s

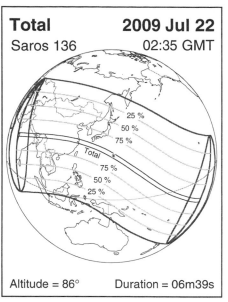

Total   **2009 Jul 22**
Saros 136   02:35 GMT

25 %
50 %
75 %
Total
75 %
50 %
25 %

Altitude = 86°   Duration = 06m39s

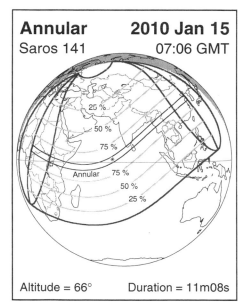

Annular   **2010 Jan 15**
Saros 141   07:06 GMT

25 %
50 %
75 %
Annular   75 %
50 %
25 %

Altitude = 66°   Duration = 11m08s

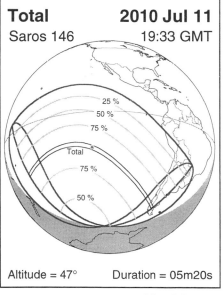

Total   **2010 Jul 11**
Saros 146   19:33 GMT

25 %
50 %
75 %
Total
75 %
50 %

Altitude = 47°   Duration = 05m20s

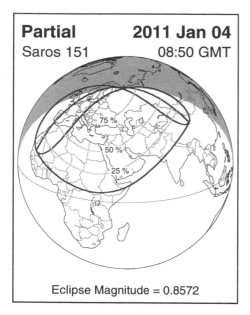

Partial   **2011 Jan 04**

Saros 151   08:50 GMT

75 %

50 %

25 %

Eclipse Magnitude = 0.8572

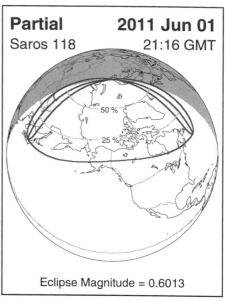

Partial   **2011 Jun 01**

Saros 118   21:16 GMT

50 %

25 %

Eclipse Magnitude = 0.6013

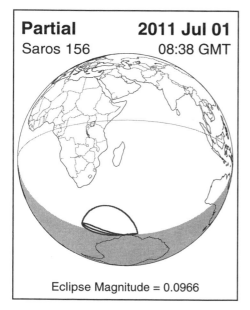

Partial   **2011 Jul 01**

Saros 156   08:38 GMT

Eclipse Magnitude = 0.0966

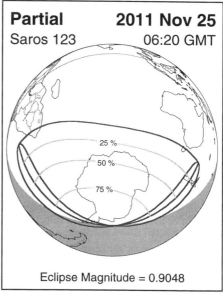

Partial   **2011 Nov 25**

Saros 123   06:20 GMT

25 %

50 %

75 %

Eclipse Magnitude = 0.9048

©1998 F. Espenak

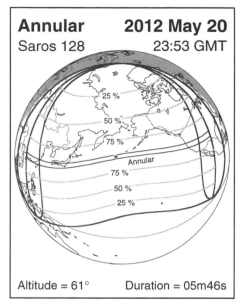

Annular   **2012 May 20**
Saros 128   23:53 GMT

Altitude = 61°   Duration = 05m46s

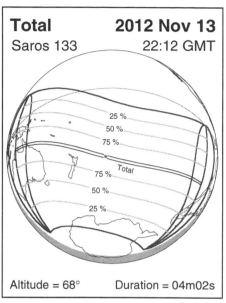

Total   **2012 Nov 13**
Saros 133   22:12 GMT

Altitude = 68°   Duration = 04m02s

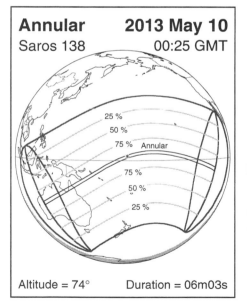

Annular   **2013 May 10**
Saros 138   00:25 GMT

Altitude = 74°   Duration = 06m03s

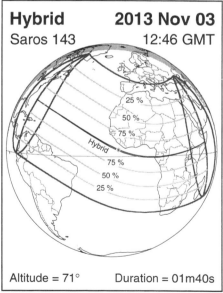

Hybrid   **2013 Nov 03**
Saros 143   12:46 GMT

Altitude = 71°   Duration = 01m40s

©1998 F. Espenak

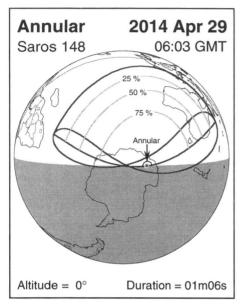

Annular	**2014 Apr 29**
Saros 148	06:03 GMT

25 %
50 %
75 %
Annular

| Altitude = 0° | Duration = 01m06s |

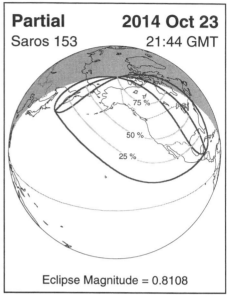

Partial	**2014 Oct 23**
Saros 153	21:44 GMT

75 %
50 %
25 %

Eclipse Magnitude = 0.8108

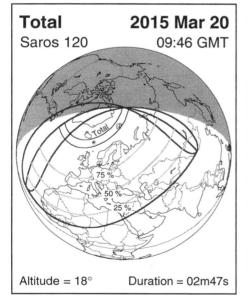

Total	**2015 Mar 20**
Saros 120	09:46 GMT

Total
75 %
50 %
25 %

| Altitude = 18° | Duration = 02m47s |

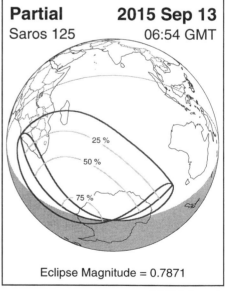

Partial	**2015 Sep 13**
Saros 125	06:54 GMT

25 %
50 %
75 %

Eclipse Magnitude = 0.7871

©1998 F. Espenak

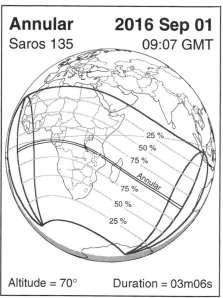

Total   **2016 Mar 09**
Saros 130   01:57 GMT

25 %
50 %
75 %
Total   75 %
50 %
25 %

Altitude = 75°   Duration = 04m09s

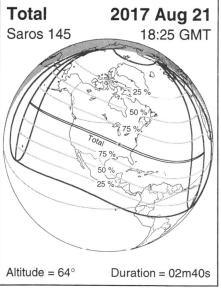

Annular   **2016 Sep 01**
Saros 135   09:07 GMT

25 %
50 %
75 %
Annular
75 %
50 %
25 %

Altitude = 70°   Duration = 03m06s

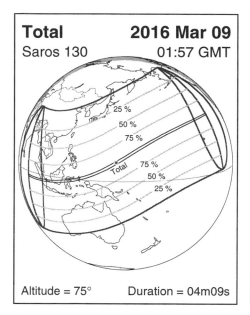

Annular   **2017 Feb 26**
Saros 140   14:53 GMT

25 %
50 %
75 %
Annular
75 %
50 %
25 %

Altitude = 63°   Duration = 00m44s

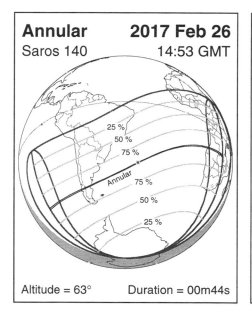

Total   **2017 Aug 21**
Saros 145   18:25 GMT

25 %
50 %
75 %
Total
75 %
50 %
25 %

Altitude = 64°   Duration = 02m40s

©1998 F. Espenak

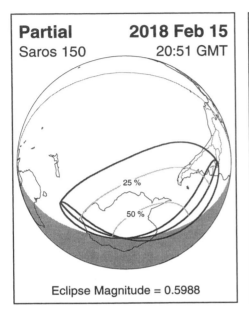

Partial   2018 Feb 15
Saros 150   20:51 GMT

25 %
50 %

Eclipse Magnitude = 0.5988

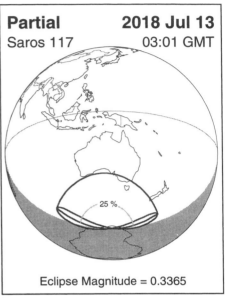

Partial   2018 Jul 13
Saros 117   03:01 GMT

25 %

Eclipse Magnitude = 0.3365

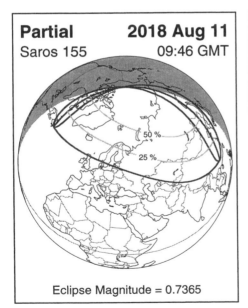

Partial   2018 Aug 11
Saros 155   09:46 GMT

50 %
25 %

Eclipse Magnitude = 0.7365

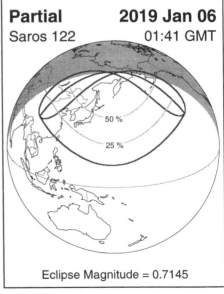

Partial   2019 Jan 06
Saros 122   01:41 GMT

50 %
25 %

Eclipse Magnitude = 0.7145

©1998 F. Espenak

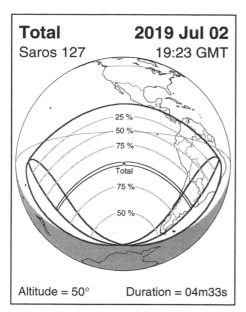

Total   **2019 Jul 02**
Saros 127   19:23 GMT

Altitude = 50°   Duration = 04m33s

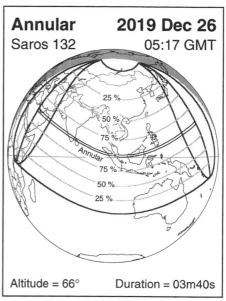

Annular   **2019 Dec 26**
Saros 132   05:17 GMT

Altitude = 66°   Duration = 03m40s

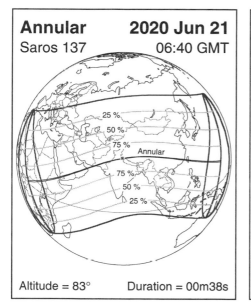

Annular   **2020 Jun 21**
Saros 137   06:40 GMT

Altitude = 83°   Duration = 00m38s

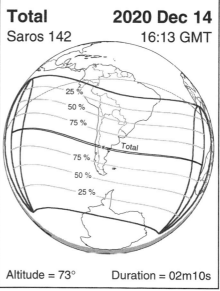

Total   **2020 Dec 14**
Saros 142   16:13 GMT

Altitude = 73°   Duration = 02m10s

©1998 F. Espenak

Central Solar Eclipses for North and South America: 2000–2020

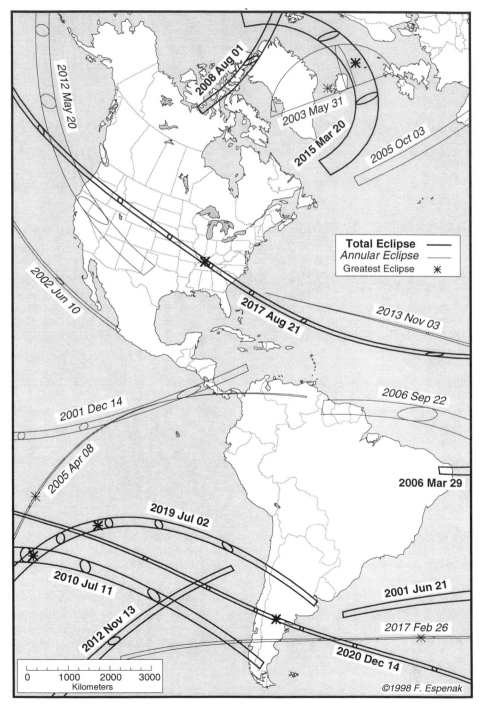

Central Solar Eclipses for Europe and Africa: 2000–2020

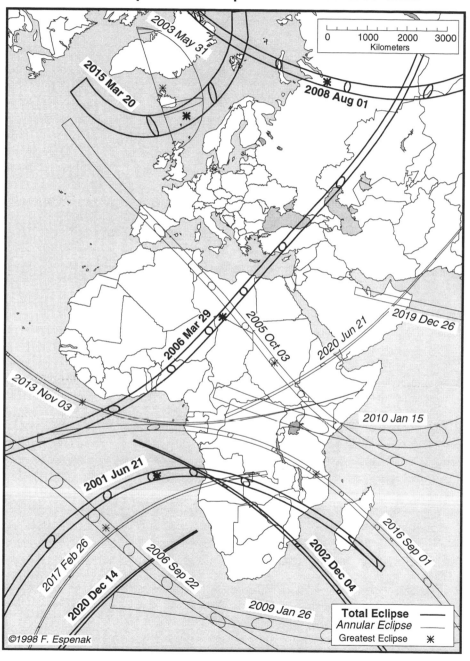

Central Solar Eclipses for Asia and Australia: 2000–2020

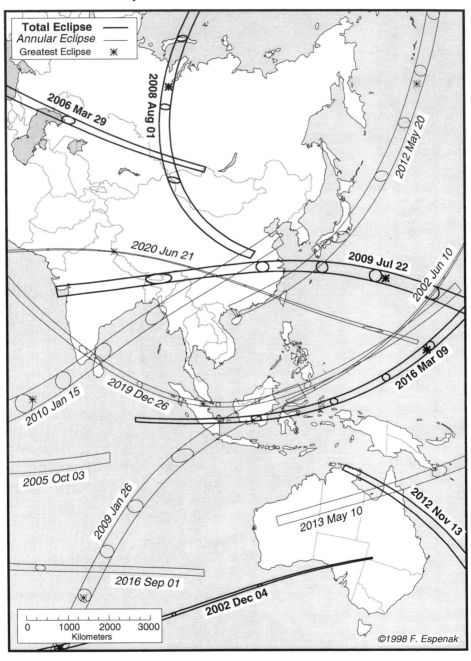

Appendix A

Total and Annular Eclipses: 1999-2052*

Maximum Duration of totality or annularity in minutes:seconds

Date	*Maximum Duration*
Path of Totality	
Date	*Maximum Duration*
Path of Annularity	

1999 August 11   2:23
Atlantic Ocean, England, France, Luxembourg, Germany, Austria, Hungary, Romania, Bulgaria, Turkey, Iraq, Iran, Pakistan, India

2001 June 21   4:57
Atlantic Ocean, Angola, Zambia, Zimbabwe, Mozambique, Madagascar

> **2001 December 14**   3:53 (annular)
> Pacific Ocean, Costa Rica, Nicaragua, Caribbean Sea

> **2002 June 10–11**   0:23 (annular)
> Pacific Ocean, touching the islands of Sangihe, Talaud, Rota, and Tinian

2002 December 4   2:04
Angola, Zambia, Botswana, Zimbabwe, South Africa, Mozambique, south Indian Ocean, southern Australia

> **2003 May 31**   3:37 (annular)
> Scotland, Iceland, Greenland (the Moon's shadow passes over the north pole of the sunward-inclined northern hemisphere, so the path of the eclipse in this unusual case moves east to west)

---

*Based on Fred Espenak: *Fifty Year Canon of Solar Eclipses: 1986–2035* (Cambridge, Massachusetts: Sky Publishing, 1987; NASA Reference Publication 1178, revised); and Hermann Mucke and Jean Meeus: *Canon of Solar Eclipses—2003 to 2526* (Vienna: Astronomisches Büro, 1983).

2003 November 23   1:57
   Antarctica

2005 April 8   0:42
   Annular-total: starts annular in western Pacific Ocean, becomes total in eastern Pacific, then becomes annular again for Costa Rica, Panama, Colombia, and Venezuela

      2005 October 3   4:31 (annular)
      Atlantic Ocean, Portugal, Spain, Algeria, Tunisia, Libya, Chad, Sudan, Ethiopia, Kenya, Somalia, Indian Ocean

2006 March 29   4:07
Eastern Brazil, Atlantic Ocean, Ghana, Togo, Benin, Nigeria, Niger, Chad, Libya, Egypt, Turkey, Russia

      2006 September 22   7:09 (annular)
      Guyana, Suriname, French Guiana, Brazil, south Atlantic Ocean

      2008 February 7   2:12 (annular)
      South Pacific Ocean, Antarctica

2008 August 1   2:27
Northern Canada, Greenland, Arctic Ocean, Russia, Mongolia, China

      2009 January 26   7:54 (annular)
      South Atlantic Ocean, Indian Ocean, Indonesia (Sumatra, Java, Borneo)

2009 July 22   6:39
India, Nepal, Bhutan, Burma, China, Pacific Ocean

      2010 January 15   11:08 (annular)
      Central Africa Republic, Congo, Uganda, Kenya, Somalia, Indian Ocean, Maldives, southernmost India, Sri Lanka, Bangladesh, Burma, China

2010 July 11   5:20
South Pacific Ocean, Easter Island, Chile, Argentina

      2012 May 20–21   5:46 (annular)
      China, Taiwan, Japan, Pacific Ocean, United States (Oregon, California, Nevada, Utah, Arizona, Colorado, New Mexico, Texas

2012 November 13   4:02
Northern Australia, south Pacific Ocean

      2013 May 9–10   6:03 (annular)
      Australia, Solomon Islands, Nauru, Pacific Ocean

2013 November 3   1:40
Annular-total: annular only at beginning and end of path: Atlantic Ocean, Gabon, Congo, Zaire, Uganda, Kenya, then annular in Ethiopia

      2014 April 29   0:00 (annular)
      Antarctica (extension of shadow cone only grazes Earth)

2015 March 20   2:47
North Atlantic Ocean, Faeroe Islands, Arctic Ocean, Svalbard

2016 March 9   4:10
Indonesia (Sumatra, Borneo, Sulawesi, Halmahera), Pacific Ocean

      2016 September 1   3:06 (annular)
      Atlantic Ocean, Gabon, Congo, Tanzania, Mozambique, Madagascar, Indian Ocean

      2017 February 26   0:44 (annular)
      Chile, Argentina, south Atlantic Ocean, Angola, Zambia, Congo

2017 August 21   **2:40**
Pacific Ocean, United States (Oregon, Idaho, Wyoming, Nebraska, Missouri, Illinois, Kentucky, Tennessee, North Carolina, South Carolina), Atlantic Ocean

2019 July 2   **4:33**
South Pacific Ocean, Chile, Argentina

>   **2019 December 26**   **3:39 (annular)**
>   Saudi Arabia, Bahrain, United Arab Emirates, Oman, southern India, Indonesia, Malaysia

>   **2020 June 21**   **0:38 (annular)**
>   Congo, Sudan, Ethiopia, Yemen, Saudi Arabia, Oman, Pakistan, northern India, China, Taiwan

2020 December 14   **2:10**
Pacific Ocean, Chile, Argentina, south Atlantic Ocean

>   **2021 June 10**   **3:51 (annular)**
>   Central Canada, north pole, Russia (Siberia)

2021 December 4   **1:55**
Antarctica

2023 April 20   **1:16**
Annular-total (total except at beginning and end of path): South Indian Ocean, western Australia, Indonesia, Pacific Ocean

>   **2023 October 14**   **5:17 (annular)**
>   United States (Oregon through Texas), Mexico (Yucatán peninsula), Central America, Colombia, Brazil

2024 April 8   **4:28**
Pacific Ocean, Mexico, United States (Texas, Oklahoma, Arkansas, Missouri, Kentucky, Illinois, Indiana, Ohio, Pennsylvania, New York, Vermont, New Hampshire, Maine), southeastern Canada, Atlantic Ocean

>   **2024 October 2**   **7:25 (annular)**
>   South Pacific Ocean, southern Chile and Argentina

>   **2026 February 17**   **2:20 (annular)**
>   Antarctica

2026 August 12   **2:19**
Greenland, Iceland, Spain

>   **2027 February 6**   **7:51 (annular)**
>   Southern Chile and Argentina, south Atlantic Ocean, Ivory Coast, Ghana, Togo, Benin, Nigeria

2027 August 2   **6:23**
Atlantic Ocean, Morocco, Spain, Algeria, Libya, Egypt, Saudi Arabia, Yemen, Somalia

>   **2028 January 26**   **10:27 (annular)**
>   Ecuador, Peru, Brazil, French Guiana, Portugal, Spain

2028 July 22   **5:10**
South Indian Ocean, Australia, New Zealand

>   **2030 June 1**   **5:21 (annular)**
>   Algeria, Tunisia, Greece, Turkey, Kazakstan, Russia, northern China, Japan

2030 November 25   **3:44**
South West Africa, Botswana, South Africa, south Indian Ocean, southeastern Australia

2031 May 21   5:26 (annular)
Angola, Zambia, Congo, Tanzania, southern India, Sri Lanka, Malaysia, Indonesia

2031 November 14   1:08
Annular-total: Pacific Ocean, Panama (total in central Pacific)

2032 May 9   0:22 (annular)
South Atlantic Ocean

2033 March 30   2:37
Alaska, Arctic Ocean

2034 March 20   4:10
Atlantic Ocean, Nigeria, Cameroon, Chad, Sudan, Egypt, Saudi Arabia, Iran, Afghanistan, Pakistan, India, China

2034 September 12   2:58 (annular)
Pacific Ocean, Chile, Bolivia, Argentina, Paraguay, Brazil, Atlantic Ocean

2035 March 9–10   0:47 (annular)
South Pacific Ocean

2035 September 2   2:54
China, Korea, Japan, Pacific Ocean

2037 July 13   3:59
Australia, New Zealand, south Pacific Ocean

2038 January 5   3:19 (annular)
Cuba, Haiti, Dominican Republic, southern Caribbean islands, Atlantic Ocean, Liberia, Ivory Coast, Ghana, Togo, Benin, Niger, Libya, Chad, Egypt

2038 July 2   1:00 (annular)
Colombia, Venezuela, Atlantic Ocean, Western Sahara, Morocco, Mauritania, Mali, Niger, Chad, Sudan, Ethiopia, Kenya

2038 December 25–26   2:18
Indian Ocean, western and southern Australia, New Zealand, Pacific Ocean

2039 June 21   4:05 (annular)
United States (Alaska), Canada, Greenland, Norway, Sweden, Finland, Estonia, Russia

2039 December 15   1:51
Antarctica

2041 April 30   1:51
South Atlantic Ocean, Angola, Congo, Uganda, Kenya, Somalia

2041 October 24–25   6:07 (annular)
Mongolia, China, Korea, Japan, Pacific Ocean

2042 April 20   4:51
Indonesia, Malaysia, Philippines, Pacific Ocean

2042 October 14   7:44 (annular)
Thailand, Malaysia, Indonesia, Australia, New Zealand

2043 April 9   0:00
Northeastern Russia (shadow cone only grazes Earth)

2043 October 3   0:00 (annular)
South Indian Ocean (extension of shadow cone only grazes Earth)

2044 February 28   2:27 (annular)
South Atlantic Ocean

2044 August 23   2:04
Greenland, Canada, United States (Montana, North Dakota)

    2045 February 16–17   **7:47 (annular)**
    New Zealand, Pacific Ocean

2045 August 12   6:06
United States (California, Nevada, Utah, Colorado, Kansas, Oklahoma, Arkansas, Mississippi, Alabama, Georgia, Florida), Haiti, Dominican Republic, Venezuela, Guyana, Suriname, French Guiana, Brazil

    2046 February 5–6   **9:42 (annular)**
    Indonesia, Papua New Guinea, Pacific Ocean, United States (California, Oregon, Nevada, Idaho)

2046 August 2   4:51
Eastern Brazil, Atlantic Ocean, Angola, Namibia, Botswana, South Africa, south Indian Ocean

    2048 June 11   **4:58 (annular)**
    United States (Kansas, Nebraska, Missouri, Iowa, Minnesota, Illinois, Wisconsin, Michigan), Canada, Greenland, Iceland, Norway, Sweden, Latvia, Estonia, Belarus, Russia, Afghanistan

2048 December 5   3:28
South Pacific Ocean, Chile, Argentina, south Atlantic Ocean, Namibia, Botswana

    2049 May 31   **4:45 (annular)**
    Peru, Colombia, Brazil, Venequela, Guyana, Suriname, Atlantic Ocean, Senegal, Mali, Guinea, Burkina, Ghana, Togo, Benin, Nigeria, Cameroon, Congo

2049 November 25   0:38
Annular-total (total through part of Indian Ocean and all of Indonesia): Saudi Arabia, Oman, Indian Ocean, Indonesia, Pacific Ocean

2050 May 20   0:21
Annular-total: South Pacific Ocean

2052 March 30   4:08
Pacific Ocean, Mexico, United States (Louisiana, Alabama, Florida, Georgia, South Carolina), Atlantic Ocean

    2052 September 22–23   **2:51 (annular)**
    Indonesia, northern Australia, south Pacific Ocean

Appendix B

---

*Based on Fred Espenak: *Fifty Year Canon of Solar Eclipses: 1986–2035* (Cambridge, Massachusetts: Sky Publishing, 1987; NASA Reference Publication 1178, revised); and Hermann Mucke and Jean Meeus: *Canon of Solar Eclipses—2003 to 2526* (Vienna: Astronomisches Büro, 1983).

1973 June 30   7:04
Guyana, Suriname, French Guiana, Atlantic Ocean, Mauritania, Mali, Alge-
ria, Niger, Chad, Central African Republic, Sudan, Uganda, Kenya, Somalia,
Indian Ocean

> **1973 December 24   12:03 (annular)**
> Costa Rica, Panama, Colombia, Venezuela, Brazil, Guyana, Atlantic Ocean,
> Mauritania, Mali, Algeria

1974 June 20   5:09
South Indian Ocean, southwestern Australia

> **1976 April 29   6:41 (annular)**
> Atlantic Ocean, Senegal, Mauritania, Mali, Algeria, Tunisia, Libya, Turkey,
> Iran, Turkmenistan, Aghanistan, India, China

1976 October 23   4:46
Tanzania, south Indian Ocean, southern Australia

> **1977 April 18   7:04 (annular)**
> South Atlantic Ocean, Namibia, Angola, Zambia, Congo, Tanzania, Indian
> Ocean

1977 October 12   2:37
Pacific Ocean, Colombia, Venezuela

1979 February 26   2:49
United States (Washington, Oregon, Idaho, Montana, North Dakota), Canada, Greenland

> **1979 August 22   6:03 (annular)**
> South Pacific Ocean, Antarctica

1980 February 16   4:08
Atlantic Ocean, Angola, Congo, Tanzania, Kenya, India, Bangladesh, Burma,
China

> **1980 August 10   3:23 (annular)**
> Pacific Ocean, Peru, Bolivia, Paraguay, Brazil

> **1981 February 4   0:33 (annular)**
> Southern Australia, south Pacific Ocean

1981 July 31   2:02
Russia, Pacific Ocean

1983 June 11   5:11
Indian Ocean, Indonesia, Papua New Guinea, Pacific Ocean

> **1983 December 4   4:01 (annular)**
> Atlantic Ocean, Gabon, Congo, Uganda, Kenya, Ethiopia, Somalia

> **1984 May 30   0:12 (annular)**
> Pacific Ocean, Mexico, United States (Louisiana, Mississippi, Alabama,
> Georgia, South Carolina, North Carolina, Virginia, Maryland), Atlantic
> Ocean, Morocco, Algeria

1984 November 22–23   2:00
Indonesia, Papua New Guinea, south Pacific Ocean

1985 November 12   1:59
South Pacific Ocean

1986 October 3   0:00.2
Annular-total: North Atlantic Ocean

1987 March 29   0:08
Annular-total: annular in Argentina; total for part of south Atlantic Ocean, Gabon, Cameroon, and part of Central African Republic; then annular in Sudan, Ethiopia, Somalia

> **1987 September 23   3:49 (annular)**
> Kazakstan, China, Mongolia, Pacific Ocean

1988 March 18   3:47
Indonesia, Philippines, Pacific Ocean

> **1988 September 11   6:57 (annular)**
> South Pacific Ocean

> **1990 January 26   2:03 (annular)**
> South Atlantic Ocean, Antarctica

1990 July 22   2:33
Finland, Russia, north Pacific Ocean

> **1991 January 15   7:53 (annular)**
> Southern Australia, New Zealand, south Pacific Ocean

1991 July 11   6:53
United States (Hawaii), Mexico, Guatemala, El Salvador, Honduras, Nicaragua, Costa Rica, Panama, Colombia, Brazil

> **1992 January 4   11:41 (annular)**
> Pacific Ocean, United States (California)

1992 June 30   5:21
Uruguay, south Atlantic Ocean

> **1994 May 10   6:14 (annular)**
> Mexico, United States (diagonally from Arizona through Missouri, New York, and Maine), southeastern Canada, Atlantic Ocean, Morocco

1994 November 3   4:23
Peru, Chile, Bolivia, Paraguay, Brazil, south Atlantic Ocean

> **1995 April 29   6:37 (annular)**
> South Pacific Ocean, Peru, Ecuador, Colombia, Brazil

1995 October 24   2:10
Iran, Afghanistan, Pakistan, India, Bangladesh, Burma, Cambodia, Vietnam, Indonesia, Pacific Ocean

1997 March 9   2:51
Mongolia, Russia, China, Arctic Ocean

1998 February 26   4:09
Pacific Ocean, Colombia, Panama, Venezuela, southern Caribbean, Atlantic Ocean

> **1998 August 22   3:14 (annular)**
> Indonesia, Malaysia, south Pacific islands

> **1999 February 16   0:40 (annular)**
> South Indian Ocean, Australia

Appendix C

Equipment Manufacturers and Retailers

Here is a brief list of some equipment manufacturers and retailers.

Camera Manufacturers

Canon U.S.A., Inc., One Canon Plaza, Lake Success, NY 11042-1198
    800-652-2666; internet: www.usa.canon.com

Minolta Corp., 101 Williams Drive, Ramsey, NJ 07446
    201-825-4000; internet: www.minolta.com

Nikon Inc., 1300 Walt Whitman Road, Melville, NY 11747-3064
    800-645-6687; internet: www.nikonusa.com

Olympus America Inc., 2 Corporate Center Drive, Melville, NY 11747-3157
    516-844-5000; internet: www.olympusamerica.com

Pentax Corp., 35 Inverness Drive East, Englewood, CO 80112
    800-877-0155; internet: www.pentax.com

Sigma Corp., 15 Fleetwood Court, Ronkonkoma, NY, 11779
    516-585-1144; internet: www.sigmaphoto.com

Camera Lens Manufacturers

Sigma Corp., 15 Fleetwood Court, Ronkonkoma, NY, 11779
    516-585-1144; internet: www.sigmaphoto.com

Tamron Industries, Inc., 125 Schmitt Blvd., Farmingdale, NY 11735
516-694-8700; internet: www.tamron.com

Tokina, THK Photo Products, Inc., 2360 Mira Mar Ave., Long Beach, CA 90815
800-421-1141; internet: www.thkphoto.com

Vivitar Corporation, 1280 Rancho Conejo Blvd., Newbury Park, CA 91320
805-498-7008; internet: www.vivitarcorp.com

Solar Filters

Abelexpress—Astronomy Division, 100 Rosslyn Rd., Carnegie, PA 15106
800-542-9001

Edwin Hirsch, 29 Lakeview Dr., Tomkins Cove, NY 10986
914-786-3738

Rainbow Symphony, 6860 Canby Ave., Unit 120, Reseda, CA 91335
818-708-8400, fax: 818-708-8470;
e-mail: 3dglasses@rainbowsymphony.com

Roger W. Tuthill, Inc., Box 1086, Mountainside, NJ 07092
908-232-1786, fax: 908-232-3804; e-mail: 75501.3622@compuserve.com

Thousand Oaks Optical, Box 5044-289, Thousand Oaks, CA 91359
805-491-3642, fax: 805-491-2393

Telescope Manufacturers

Astro-Physics, Inc., 11250 Forest Hills Rd., Rockford, IL 61115
815-282-1513, fax: 815-282-9847

Celestron International, 2835 Columbia St., Torrance, CA 90503, USA
310-328-9560, fax: 310-212-5835

Meade Instruments Corporation, 6001 Oak Canyon, Irvine, CA 92620
714-451-1450, fax: 714-451-1460

Questar Corp., 6204 Ingham Rd., New Hope, PA 18938
215-862-5277, fax: 215-862-0512; e-mail: questar@erols.com

Takahashi, c/o Land, Sea & Sky Unlimited, 3110 S. Shepherd, Houston, TX 77098
713-529-3551, fax: 713-529-3108

Tasco, 7600 Northwest 26th St., Miami, FL 33122
305-591-3670

Tele Vue Optics, Inc., 100 Route 59, Suffern, NY 10901
914-357-9522, fax: 914-357-9523

Unitron Inc., P.O. Box 469, Bohemia, NY 11716
516-589-6666, fax: 516-589-6975; e-mail: info@unitronusa.com

Telescope and Accessories Retailers

Adorama Camera, 42 West 18th St., New York, NY 10011
    212-741-0466, fax: 212-463-7223, e-mail: goadorama@aol.com

Astronomics, 2401 Tee Circle, Suite 105, Norman, OK 73069
    405-364-0858, fax: 405-447-3337; e-mail: questions@astronomics.com

Edwin Hirsch, 8740 Egret Isle Terrace, Lake Worth, FL 33467
    phone/fax: 407-641-2851

F. C. Meichsner Co., 182 Lincoln St., Boston, MA 02111
    617-426-7092, fax: 617-426-0837

Focus Camera Inc., 4419-21 13th Ave., Brooklyn, NY 11219
    718-437-8810, fax: 718-437-8811; e-mail: orders@focuscamera.com

Khan Scope Centre, 3243 Dufferin St., Toronto, ON M6A 2T2, Canada
    416-783-4140, fax: 416-783-7697; e-mail: khan@khanscope.com

Lumicon, 2111 Research Dr. #5, Livermore, CA 94550
    925-447-9583, fax: 925-447-9589

Orion Telescopes & Binoculars, P.O. Box 1815, Santa Cruz, CA 95061
    831-763-7000, fax: 831-763-7017; e-mail: sales@oriontel.com

Pocono Mountain Optics, 104 N P 502 Plaza, Moscow, PA 18444
    717-842-1500, fax: 717-842-8364; e-mail: pocmtnop@ptdprolog.net

Scope City, 730 Easy St., Simi Valley, CA 93065
    805-522-6646, fax: 805-582-0292

Shutan Camera & Video, 312 W. Randolph, Chicago, IL 60606
    312-332-2000, fax: 312-332-2019

Swift Instruments, 952 Dorchester Ave., Boston, MA 02125
    617-436-2960, fax: 617-436-3232

Wholesale Optics, 59 Mine Hill Rd., New Milford, CT 06776
    860-355-3132; e-mail: astroptx@astroptx.com

Appendix D

Eclipse and Astronomy Software

Here is a brief list of astronomical software publishers offering programs that can perform some level of eclipse prediction. All programs are IBM compatible unless otherwise noted.

Commercial Software

ARC Science Simulations: *Dance of the Planets*
    P.O. Box 1955, Loveland, CO 80539
    970-663-3223; e-mail: arcmail@arcinc.com

Carina Software: *Voyager II* (Macintosh & IBM)
    12919 Alcosta Blvd., Suite 7, San Ramon, CA 94583
    510-355-1266; fax: 510-355-1268; e-mail: support@carinasoft.com

CEB Metasystems Inc.
    1200 Lawrence Dr., #175, Newbury Park, CA 91320
    800-232-7830; fax: 805-498-5987

Expert Software
    800 Douglas Rd., North Tower, Suite 750, Coral Gables, FL 33134
    305-567-9990; fax: 305-443-0786

Maris Multimedia Inc.: *Red Shift 3* (Macintosh & IBM)
    4040 Civic Center Dr., Suite 200, San Rafael, CA 94903
    415-492-2819; e-mail: redshift@maris.com

Personal MicroCosms
   8547 E. Arapahoe Road, Suite J-147, Greenwood Village, CO 80112
   303-753-3268

Project Pluto: *The Guide 6.0*
   168 Ridge Rd., Bowdoinham, ME 04008
   207-666-5750; e-mail: b.gray2@genie.geis.com

Sienna Software: *Starry Night Deluxe 2* (Macintosh & IBM)
   Suite 303, 411 Richmond St. E., Toronto, ON M5A 3S5, Canada
   416-410-0259; fax: 416-410-0359; e-mail: contact@siennasoft.com

Software Bisque: *The Sky*
   912 12th Street, Suite A, Golden, CO 80401
   303-278-4478; fax: 303-279-1180; e-mail: smb@bisque.com

Zephyr Services: *Sun Tracker Plus, Total Eclipse, Eclipse Complete*
   1900 Murray Ave., Pittsburgh, PA 15217
   800-533-6666; fax: 412-422-9930; e-mail: mail@zephyrs.com

Shareware and Freeware

Solar & Lunar Eclipse Simulator (DOS)
   internet: www.astronomy.ch/

Solar 1.4
   Matthew M. Merrill, 8927 Virginia Ave. Apt. B, South Gate, CA 90280
   213-567-1088

Eclipse v5.01
   Don Nicholson, 2124 Linda Flora Dr., Los Angeles, CA 90077
   310-476-4413

Planetarium with Lunar Eclipse Simulator
   Christian Nuesch, Haldenstrasse 12, CH-8320 Fehraltorf, Switzerland
   e-mail: nuesch@active.ch

Astro Lab 2 (32-bit DOS shareware)
   Bob Denton, GE Capital IT Solutions, Melbourne, Victoria, Australia
   +61-3-9541-5751; fax: +61-3-9544-6295;
   e-mail: Bob.Denton@computer.org

Software Reviews

John Mosley's Astro Software Revue
   internet: www.skypub.com/software/mosley.html

Book for Writing Eclipse Software

Jean Meeus: *Elements of Solar Eclipses: 1951–2200* (1989)
   Willmann-Bell, Inc., P.O. Box 35025, Richmond, VA 23235
   804-320-7016; fax: 804-272-5920
   Detailed explanation, equations, and examples of how to predict solar
eclipses.

Appendix E

Internet Resources

Here is a brief list of eclipse-related resources available through the Internet on the World Wide Web. Links to all these sites and more are available through:

> *sunearth.gsfc.nasa.gov/eclipse/toplink.html*

Solar Eclipses—General Resources

Astronomical Pictures & Animations—
> *graffiti.u-bordeaux.fr/MAPBX/roussel/astro.english.html*
> Frank Roussel's page has a number of eclipse maps with information.

Earth View Eclipse Network—
> *www.earthview.com/*
> Bryan Brewer (author of *Eclipse*) offers tutorials and information on future solar eclipses.

Eclipse!—
> *ourworld.compuserve.com/homepages/pharrington/sw13.htm*
> Phil Harrington (author of *Eclipse!*) gives information on future eclipses.

Eclipse Chaser Home Page—
> *www.eclipsechaser.com/*
> Jeffrey Charles' solar eclipse travelogs, information, and images.

The Eclipse Zone—
> *www.geocities.com/CapeCanaveral/7137/ez.htm*
> Information and graphics on eclipses in the near future (Worachate Booplod).

Fred Espenak's Eclipses Home Page—
    sunearth.gsfc.nasa.gov/eclipse/eclipse.html
    Information on 5,000 years of solar and lunar eclipses, photography,
    observer reports, and more.

Fred Espenak's 1999 Total Solar Eclipse Page—
    sunearth.gsfc.nasa.gov/eclipse/TSE1999/TSE1999.html
    Detailed maps, tables, and weather for the 1999 total solar eclipse.

GSFC Solar Data Analysis Center's Eclipse Information Page—
    umbra.nascom.nasa.gov/eclipse/
    Official site for NASA's Solar Eclipse Bulletins by Fred Espenak and Jay
    Anderson, with information on how to order the bulletins.

IAU Eclipse Home Page—
    www.williams.edu/Astronomy/IAU_eclipses/
    Professor Jay Pasachoff's Internaional Astronomical Union Eclipse Page
    describes scientific eclipse experiments.

IOTA Occultations Page—
    www.sky.net/~robinson/iotandx.htm
    The International Occultation Timing Association home page features
    predictions for lunar occultations and eclipses.

Sky Online's Eclipse Page—
    www.skypub.com/eclipses/eclipses.shtml
    Sky Publishing Corporation's Eclipse Page provides extensive
    information, much of it adapted from the pages of *Sky & Telescope*
    magazine.

Solar Eclipse Filters

Rainbow Symphony Eclipse Shades—
    www.rainbowsymphony.com/soleclipse.html

Solar Eclipse Photography

Wendy Carlos Eclipse Page—
    www.wendycarlos.com/eclipse.html
    Features eclipse-chaser Wendy Carlos' eclipse composite photography.

High Moon—
    eclipse.span.ch/
    Olivier Staiger's unusual take on eclipses and astronomical events.

Dale Ireland's Eclipse Page—
    www.drdale.com/eclipses/
    Dale Ireland offers a gallery of both solar and lunar eclipse photographs.

Solar Eclipse Composites—
    www.alliancerevolution.com/astro/
    Eclipse-chaser Benjamin Gomes-Casseres' digital eclipse composites.

Numazawa's Eclipse Page—
> *www1.nisiq.net/~numazawa/st/eclipimg.html*
> Astrophotographer Shigemi Numazawa's formula for digital eclipse composites.

Bob Yen's Eclipses—
> *www.comet-track.com/eclipse/secl.html*
> Astrophotographer Bob Yen's gallery of offbeat eclipse photos.

Solar Eclipse Mailing Lists

Earth View Eclipse Mailing List—
> *www.earthview.com/resources/mailinglist.htm*
> From Bryan Brewer, author of *Eclipse*.

Solar Eclipse Mailing List
> For instructions, e-mail Patrick Poitevin: *PPoitevin@village.uunet.be*

Lunar Eclipses

Tony Mallama's Lunar Eclipse Page—
> *www.serve.com/meteors/mallama/lunarecl/index.html*
> Lunar eclipse models predicting apparent visual magnitudes.

Lunar Eclipse Observer—
> *www-clients.spirit.net.au/~minnah/LEOx.html*
> Lunar eclipse observing, Caldwell Lunar Observatory, Australia.

Astronomy Resources

Amateur Astronomy Clubs—
> *www.aspsky.org/html/resources/amateur.html*
> Need to know where the nearest astronomy club is? Find it here.

Astronomy Net—
> *www.astronomy.net/*
> Features forums; links a searchable astronomy database.

Astronomy Mall—
> *astronomy-mall.com/*
> Links to many astronomical products, services, and free classified listings for amateur astronomers.

Yahoo Astronomy Links—
> *www.yahoo.com/Science/Astronomy/*
> All kinds of links related to astronomy.

Planetariums—
> *www.aspsky.org/html/resources/planetarium.html*
> List of planetariums around the world.

Astronomy Magazine—
> *www.kalmbach.com/astro/*
> *Astronomy's* home page offers news, sky events, and product information.

Astronomy Now Online—
  www.astronomynow.com
  Astronomy Now magazine's page features news, night sky, and astro-listings.

Sky Online—
  www.skypub.com/
  Sky & Telescope's page furnishes articles, news, sky events, and links.

The Astronomy Cafe—
  www2.ari.net/home/odenwald/cafe.html
  Sten Odenwald's web site for the "astronomically disadvantaged."

Abrams Planetarium Skywatcher's Diary—
  www.pa.msu.edu/abrams/diary.html
  Monthly guide to what's happening in the night sky.

Earth & Sky Online—
  www.earthsky.com
  Daily guide to what's happening in the night sky.

Twelve-Year Planetary Ephemeris: 1995-2006—
  sunearth.gsfc.nasa.gov/eclipse/type/type.html
  Fred Espenak's tables of planet, Sun, and Moon positions, dates of Moon phases, and planetary alignments.

Mike Boschat's Astronomy Page—
  www.atm.dal.ca/~andromed/
  Boschat maintains a comprehensive site of astronomy links.

Used Telescopes (classified ads)

AstroMart—
  www.astromart.com/class.html

The Starry Messenger—
  ww1.starrymessenger.com/webpages/tsm

Eclipse Tours and Expeditions

Astronomical Tours—
  www.AstronomicalTours.com/

Explorer Tours—
  www.star.ucl.ac.uk/~hwm/explorer.htm

Hole in the Sky Tours—
  www.holeinthesky.com/

S&T and Scientific Expeditions—
  www.skypub.com/market/scienx/scienx.html

Wilderness Travel—
  www.wildernesstravel.com

Travel Advisories

Travel Warnings—
> *travel.state.gov/travel_warnings.html*
> Travel alerts issued by the U.S. State Department for American citizens traveling abroad.

Maps

Excite World Maps—
> *www.city.net/maps/*
> Online maps of every state and country in the world.

Microsoft Terra Server—
> *terraserver.microsoft.com/*
> Satellite imagery for locations around the world.

U.S. Geological Survey—
> *info.er.usgs.gov/*
> Complete topographic maps of the United States.

World of Maps—
> *www.worldofmaps.com/*
> Online map store for maps of any part of the world.

Weather Information

National Weather Service—
> *205.156.54.206/*
> NOAA's National Weather Service home page.

WeatherNet—
> *cirrus.sprl.umich.edu/wxnet/*
> WeatherNet is one of the the Internet's principal sources of weather information.

The Weather Underground—
> *www.wunderground.com/*
> University of Michigan's world forecasts.

The Weather Channel—
> *www.weather.com*
> The Weather Channel's extensive site for U.S.A. and world weather forecasts.

IntelliCast Weather—
> *www.intellicast.com/weather/intl/*
> Latest weather forecasts around the world.

Appendix F

Popular Astronomy Magazines and Journals

(English language)

Astronomy
P.O. Box 1612, Waukesha, Wisconsin 53187, United States
414-796-8776; fax: 414-798-6468; e-mail: astro@astronomy.com

Astronomy Now
P.O. Box 175, Tonbridge, Kent TN10 4ZY, United Kingdom
01903-266165; fax: 01732-356230; e-mail:
editorial@astronow.demon.co.uk

Griffith Observer
Griffith Observatory, 2800 East Observatory Road, Los Angeles, California
90027, United States
213-664-1181; fax: 213-663-4323; e-mail: GriffithObs.org

Journal of the British Astronomical Association
Burlington House, Piccadilly, London W1V 9AG, United Kingdom
0171-734-4145; fax: 0171-439-4629; e-mail: office@baahq.demon.co.uk

Mercury
Astronomical Society of the Pacific, 390 Ashton Avenue, San Francisco,
California 94112, United States
415-337-1100; fax: 415-337-5205; e-mail: mercury@aspsky.org

Sky & Telescope
P.O. Box 9111, Belmont, Massachusetts 02178-9111, United States
617-864-7360; fax: 617-576-0336; e-mail: www.skypub.com/email.html

StarDate
McDonald Observatory, University of Texas, Austin, Texas 78712,
United States
512-471-5285; fax: 512-471-5060

Appendix G

NASA Solar Eclipse Bulletins

NASA astronomer Fred Espenak and Canadian meteorologist Jay Anderson publish special NASA bulletins for each major eclipse of the Sun. Produced through NASA's Reference Publication (RP) series, the eclipse bulletins are prepared in cooperation with the Working Group on Eclipses of the International Astronomical Union and are provided as a public service to both the professional and lay communities, including educators and the media. Each eclipse bulletin is a complete reference for a specific eclipse and contains detailed predictions, tables, maps, and weather prospects.

To date, the following eclipse bulletins have been published:

Annular Solar Eclipse of 1994 May 10 (NASA RP 1301)
Total Solar Eclipse of 1994 November 03 (NASA RP 1318)
Total Solar Eclipse of 1995 October 24 (NASA RP 1344)
Total Solar Eclipse of 1997 March 09 (NASA RP 1369)
Total Solar Eclipse of 1998 February 26 (NASA RP 1383)
Total Solar Eclipse of 1999 August 11 (NASA RP 1398)

Future bulletins currently planned include the eclipses of 2001 June 21; 2002 December 4; 2003 May 31; 2003 November 23; and 2006 March 29.

Single copies of the bulletins are available at no cost by sending a 9-by-12-inch self-addressed envelope stamped with sufficient postage for 11 ounces (310 grams) and with the eclipse date printed in the lower left corner. Please send stamps only. Cash or checks cannot be accepted. Requesters from outside the United States and Canada should send nine international postal coupons. An order form for the bulletins and the current availability of past and future eclipse bulletins can be found at:

http://sunearth.gsfc.nasa.gov/eclipse/RPrequest.html

or write directly to:

Fred Espenak
NASA/Goddard Space Flight Center
Code 693
Greenbelt, Maryland 20771
USA
e-mail: espenak@gsfc.nasa.gov

The NASA eclipse bulletins are also available on-line via NASA/Goddard's Solar Data Analysis Center eclipse page:

http://umbra.nascom.nasa.gov/eclipse/

Appendix H

Chronology of Discoveries about the Sun*

Year
Discoveries about the Sun Made During Solar Eclipses
> *Year*
> *Other Discoveries about the Sun*

2159 to 1948 b.c.
Legendary dates from China in the *Shu Ching* of the first recorded solar eclipse. In his myth, Chinese astronomers Hsi and Ho fail to prevent or predict or properly react to an eclipse and are ordered to be executed by an angry emperor

1307 b.c.
First recorded observation of the corona (or prominences?) during a solar eclipse—in China on oracle bones: "three flames ate up the Sun, and a great star was visible"

1223 b.c., March 5
Oldest record of a verifiable solar eclipse—on a clay tablet in the ruins of Ugarit (Syria)

1217 b.c.
Oldest Chinese record of a verifiable solar eclipse—inscriptions on an oracle bone by the Shang people

585 b.c., May 28
A total eclipse in the midst of a battle between the Lydians and Medes scares both sides; hostilities are suspended, according to the Greek historian Herodotus (several other dates are possible)

---

*With advice from Alan Clark, Joseph Hollweg, Charles Lindsey, and Jay Pasachoff.

6th century b.c.
Babylonians (Chaldeans) are said to be able to predict eclipses of the Sun
and Moon, supposedly based on cycles such as the saros, but more likely
from the positions of the Moon's nodes

450 b.c.?
Anaxagoras (Greece) is probably the first to realize that the Moon is illuminated by the Sun, thus providing a scientific explanation of eclipses

> **450 b.c.?**
> Greek philosopher Anaxagoras proposes that the Sun is a giant glowing
> ball of rock; he is banished from Athens for blasphemy

> **431 b.c., August 3**
> Oldest European record of a verifiable solar eclipse (annular)—by the
> Greek historian Thucydides

> **330 b.c.?**
> Aristotle (Greece) proposes that the Sun is a sphere of pure fire, unchang-
> ing and without imperfections

130 b.c.
Greek astronomer Hipparchus uses the position of the Moon's shadow during a solar eclipse to estimate the distance to the Moon (accurate to about
13%)

20 b.c.?
Liu Hsiang first in China to explain that the Moon's motion hides the Sun
to cause solar eclipses

a.d. 150?
Ptolemy (Alexandria) demonstrates the computation of solar and lunar
eclipses based on their apparent motions rather than the periodic repetition
of eclipses

334, July 17
Firmicus (Sicily) is first to report solar prominences, seen during an annular
eclipse

418, July 19
First report of a comet discovered during a solar eclipse, seen by the historian Philostorgius in Asia Minor

9th century
Shadow bands during a total eclipse are described for the first time—in the
Volospa, part of the old German poetic edda

968, December 22
First clear description of the corona seen during a total eclipse—by a chronicler in Constantinople

1605
Johannes Kepler (Germany) is the first to comment scientifically on the
solar corona, suggesting that it is light reflected from matter around the Sun
(based on reports of eclipses; he never saw a total eclipse)

> **1609**
> Galileo Galilei (Italy) and Johannes Fabricius (Holland) independently
> observe sunspots to be part of the Sun, disproving Aristotle's contention
> that the Sun is unchanging and free of "imperfections"

1683
Gian Domenico Cassini (Italy/France) explains zodiacal light as sunlight reflected from small solid particles in the plane of the solar system

1687
Isaac Newton (England) publishes his Principia, including the law of universal gravitation, which makes precise long-range eclipse prediction possible

1695
Edmond Halley (England) is first to notice that the reported times and places of ancient eclipses do not correlate with calculations backward from his era; he concludes correctly that the Moon's orbit has changed slightly (secular acceleration)

1706, May 12
An English ship captain named Stannyan, on vacation in Switzerland, reports a reddish streak (chromosphere? prominence?) along the rim of the Sun as the eclipse becomes total

1715, May 3
Edmond Halley (England), during an eclipse in England, is the first to report the phenomenon later known as Baily's Beads; also notes bright red prominences and the east-west asymmetry in the corona, which he attributes to an atmosphere on the Moon or Sun

1715
Gian Domenico Cassini (Italy/France) proposes that the light responsible for the corona also causes the zodiacal light

1724, May 22
Giacomo Filippo Maraldi (Italy/France) concludes that the corona is part of the Sun because the Moon traverses the corona during an eclipse

1733, May 13
Birger Wassenius (Sweden), observing an eclipse near Göteborg, is the first to report prominences visible to the unaided eye; he attributes them to the Moon

1749
Richard Dunthorne (United Kingdom) calculates the approximate secular acceleration of the Moon based on Halley's 1695 findings

1774
Alexander Wilson (Scotland) shows that sunspots are depressions in the Sun's photosphere, not clouds above it or mountains or volcanic deposits on it

1795
William Herschel (United Kingdom) proposes that sunspots are holes in the Sun's hot clouds through which the dark, cool, solid surface of the Sun can be seen; he also suggests that this surface may be inhabited

1800
William Herschel (United Kingdom) founds science of solar physics by measuring the temperature of various colors in the Sun's spectrum; he detects infrared radiation beyond the visible spectrum

1801
Johann Wilhelm Ritter (Germany), following the lead of Herschel, uses the spectrum of the Sun to establish the existence of ultraviolet radiation

1802
Dark (absorption) lines are discovered in the Sun's spectrum by William Wollaston (United Kingdom)

1806, June 16
José Joaquin de Ferrer (Spain), observing at Kinderhook, New York, gives the name *corona* to the glow of the faint outer atmosphere of the Sun seen during a total eclipse; he proposes that the corona must belong to the Sun, not the Moon, because of its great size

1817
Joseph Fraunhofer (Germany) independently discovers and catalogs many hundreds of dark lines in the Sun's spectrum

1820
Carl Wolfgang Benjamin Goldschmidt (Germany) calls attention to the shadow bands visible just before and after totality at some eclipses (based on the eclipse of November 19, 1816?)

1824
Friedrich Bessel (Prussia) introduces an easier way (Besselian elements) to make eclipse predictions

1833
David Brewster (United Kingdom) shows that some of the dark lines in the Sun's spectrum are due to absorption in the Earth's atmosphere, but most are intrinsic to the Sun

1836, May 15
Francis Baily (United Kingdom), during an annular eclipse in Scotland, calls attention to the brief bright beads of light that appear close to totality as the Sun's disk is blocked except for sunlight streaming through lunar valleys along the limb. This phenomenon becomes known as *Baily's Beads*

1837–1838
John F. W. Herschel (United Kingdom) and Claude Servais Mathias Pouillet (France) independently make the first quantitative measurements of the heat emitted by the Sun (about half the actual value). James David Forbes (United Kingdom) obtains a more accurate value in 1842, but it is considered less reliable

1842, July 8
Francis Baily (United Kingdom), at an eclipse in Italy, focuses attention on the corona and prominences and identifies them as part of the Sun's atmosphere

1843
Heinrich Schwabe (Germany), after a 17-year study, discovers the sunspot cycle of 10 years (now known to average 11 years)

1845
First clear photograph of the Sun: a daguerreotype by Hippolyte Fizeau and Léon Foucault (France)

1848
Julius Robert Mayer (Germany) shows through calculations that the Sun cannot shine a significant length of time by ordinary chemical burning; he incorrectly attributes the energy of the Sun to the heat released by the impact of meteoroids on the Sun

1851, July 28
First astronomical photograph of a total eclipse: a daguerreotype by Berkowski at Königsberg, Prussia

1851, July 28
Robert Grant and William Swan (United Kingdom) and Karl Ludwig von Littrow (Austria) determine that prominences are part of the Sun because the Moon is seen to cover and uncover them as it moves in front of the Sun

1851, July 28
George B. Airy (United Kingdom) is the first to describe the Sun's chromosphere: he calls it the *sierra*, thinking that he is seeing mountains on the Sun, but he is actually seeing small prominences (spicules) that give the chromosphere a jagged appearance. Because of its reddish color, J. Norman Lockyer names this layer of the Sun's atmosphere the *chromosphere* in 1868

1852
Johann von Lamont (Germany), after 15 years of observations, discovers a 10.3-year activity cycle in the Earth's magnetic field, but does not correlate it with the sunspot cycle

1852
Edward Sabine (Ireland/United Kingdom), Johann Rudolf Wolf (Germany), and Alfrede Gautier (France) independently link the sunspot cycle to magnetic fluctuations on Earth; the study of solar-terrestrial relationships begins

1854
Hermann von Helmholtz (Germany) attributes the energy of the Sun to heat from gravitational contraction

1858
Richard C. Carrington (United Kingdom), studying sunspots, identifies the Sun's axis of rotation and discovers that the Sun's rotational period varies with latitude. He also discovers that the latitude of sunspots migrates toward the equator during the course of a sunspot cycle

1859, September 1
Richard C. Carrington and R. Hodgson (United Kingdom) are the first to observe a flare on the Sun. They both also note that a magnetic storm in progress on Earth intensifies soon afterward, but they refrain from connecting the two events

1859
Gustav Kirchhoff (Germany) uses spectroscopy to show that the surface of the Sun cannot be solid and that the Sun's atmosphere (which he identifies with the corona) is responsible for the dark lines in the Sun's spectrum

1860, July 18
First wet plate photographs of an eclipse; they require 1/30 of the exposure time of a daguerreotype

1860, July 18
Warren De La Rue (United Kingdom) and Angelo Secchi (Italy) use photography during a solar eclipse in Spain to demonstrate that prominences (and hence at least that region of the corona) are part of the Sun, not light scattered by the Earth's atmosphere or the edge of the Moon, because the corona looks the same from sites 250 miles apart

1859-1861
Gustav Kirchhoff and (in part) Robert Bunsen (Germany) identify 12 elements in the Sun based on lines in its spectrum

1861-1862
Gustav Kirchhoff (Germany) maps the solar spectrum

1868, August 18
During an eclipse seen from the Red Sea through India to Malaysia and New Guinea, prominences are first studied with spectroscopes and shown to be composed primarily of hydrogen by James Francis Tennant (United Kingdom), John Herschel (United Kingdom—son of John F. W. Herschel, grandson of William), Jules Janssen (France), Georges Rayet (France), and Norman Pogson (United Kingdom/India)

1868
Pierre Jules César Janssen (France) and J. Norman Lockyer (United Kingdom) independently demonstrate that prominences are part of the Sun (not Moon) by observing them days after the eclipse of August 18

1868
J. Norman Lockyer (United Kingdom) identifies a yellow spectral line in the Sun's corona as the signature of a chemical element as yet unknown on Earth. He later names it *helium*, after the Greek word *helios*, the Sun. Helium is first identified on Earth by William Ramsay in 1895

1869, August 7
Charles Augustus Young and William Harkness (United States) independently discover a new bright (emission) line in the spectrum of the Sun's corona, never before observed on Earth; they ascribe it to a new element and it is named *coronium*. In 1941, this green line is identified by Bengt Edlén (Sweden) as iron that has lost 13 electrons

1869
Anders Jonas Angström (Sweden) produces an improved map of the solar spectrum (using a diffraction grating rather than a prism) and introduces angstrom unit for measuring wavelength

1869
Thomas Andrews (Ireland) shows more conclusively than those before him that the Sun must be made essentially of hot gas

1870, December 2
Jules Janssen (France) uses a balloon to escape the German siege of Paris in order to study the December 22 eclipse in Algeria. He reaches Algeria, but the eclipse is clouded out

1870, December 22
Charles A. Young (United States), observing an eclipse in Spain, discovers that the chromosphere is the layer in the solar atmosphere that produces the dark lines in the Sun's spectrum

1870
Samuel P. Langley (United States) uses the Doppler effect of the Sun's rotation to show that it displaces the dark (absorption) lines in the solar spectrum

1871, December 12
Jules Janssen (France) uses spectroscopy from an eclipse in India to propose that the corona consists of both hot gases and cooler particles and hence is part of the Sun

1872, August 3
Charles A. Young (United States) observes a flare on the Sun with a spectroscope; he calls attention to its coincidence with a magnetic storm on Earth

1874
Samuel P. Langley (United States) proposes that the bright granules in the Sun's photosphere are columns of hot gas rising from the interior and the dark interstices are cooler gases descending; he also proposes that the bright granules are responsible for almost all of the Sun's light

1871/1878
Jules Janssen (France) notices that the shape of the corona changes with the sunspot cycle. At sunspot maximum, the corona is rounder (1871); at sunspot minimum, the corona is more equatorial (1878). This discovery is the most convincing evidence that the corona is part of the Sun

1878, July 29
Height of search for intra-Mercurial planet Vulcan using eclipses to block the Sun. Several observers claim sightings, but they were never confirmed. The problem is finally resolved by Einstein in his general theory of relativity in 1916

1878, July 29
Samuel P. Langley and Cleveland Abbe (United States), observing from Pike's Peak in Colorado, and Simon Newcomb (United States), observing from Wyoming, notice coronal streamers extending more than 6 degrees from the Sun along the ecliptic and suggest that this glow is the origin of the zodiacal light

1882, May 17
A comet is discovered and photographed by Arthur Schuster (Germany/United Kingdom) during an eclipse in Egypt: the first time a comet discovered in this way has been photographed

1884
Marie Alfred Cornu (France) applies Langley's work using the Doppler effect of the Sun's rotation to distinguish between dark (absorption) lines created in the Sun's atmosphere and those created in the Earth's atmosphere

1887
Theodor von Oppolzer's (Czechoslovakia) monumental *Canon of Eclipses* published, giving details of almost all solar and lunar eclipses from 1207 B.C. to A.D. 2161.

1887, August 19
Dmitry Ivanovich Mendeleev (Russia) uses a balloon to ascend above the cloud cover to an altitude of 11,500 feet (3.5 kilometers) to observe an eclipse in Russia

1889
Henry Augustus Rowland (United States) produces an essentially modern map of the solar spectrum, identifying a total of 36 elements present in the Sun

1893 & 1894
George Ellery Hale (United States) and Henri Deslandres independently develop spectroheliographs to photograph the Sun's chromosphere, prominences, and flares without waiting for eclipses

1894
William E. Wilson and P. L. Gray (Ireland) are the first to measure with reasonable accuracy the effective temperature of the Sun's photosphere: 11,200°F (6,200°C), about 800°F (400°C) too high

1899
Friedrich Ginzel (Austria), Oppolzer's coworker, uses data from the *Canon of Eclipses* for his *Special Canon of Solar and Lunar Eclipses*, which lists references in classical literature to eclipses between 900 B.C. and A.D. 600.

1904
George Ellery Hale (United States) establishes on Mt. Wilson in California the first large solar observatory

1908
George Ellery Hale (United States) shows that sunspots are regions with strong magnetic fields

1911
Albert Einstein (Germany), working on his general theory of relativity, proposes that gravity bends light and that this phenomenon might be observed during a solar eclipse

1916
Einstein publishes his complete general theory of relativity with a revised prediction for the gravitational deflection of starlight

1919, May 29
Arthur S. Eddington (United Kingdom) and coworkers, observing a total solar eclipse from Principe and Brazil, confirm the bending of starlight by gravity as predicted by Einstein in his general theory of relativity

1922, September 21
William Wallace Campbell and Robert J. Trumpler (United States) reconfirm Einstein's relativistic bending of starlight during an eclipse in Wallal, Australia

1926
Arthur Eddington (United Kingdom) proposes that the Sun and stars derive their energy from nuclear reactions at their core

1930
Bernard Lyot (France) invents the coronagraph, which creates an artificial

eclipse inside a telescope so that the corona can be studied outside of eclipses

1932, August 31
G. G. Cillié (United Kingdom) and Donald H. Menzel (United States) use eclipse spectra to show that the Sun's corona has a higher temperature (faster atomic motion) than the photosphere. Confirmed, with much higher temperatures, by R. O. Redman during an eclipse in South Africa on October 1, 1940

1951
Ludwig F. Biermann (Germany) discovers the solar wind, a stream of charged gas particles continuously ejected from the Sun in all directions, based on observations of charged gases in comet tails

1952
Discovery of strong absorption lines of the molecule carbon monoxide in the solar infrared spectrum by Leo Goldberg and colleagues (United States)

1958
Eugene Parker (United States) provides a theoretical model for the solar wind by showing that the corona must be expanding and by demonstrating how to calculate flow speeds and densities

1959
Robert Leighton, Robert Noyes, and George Simon (United States) discover 5-minute oscillations of the surface of the Sun, leading to the birth of helioseismology and the "acoustic" exploration of the Sun's interior

1961
Marcia Neugebauer and her coworkers (United States) use NASA's Mariner 2 spacecraft to confirm major features of Eugene Parker's 1958 model of the solar wind

1962–1972
NASA's Orbiting Solar Observatories (OSOs)

1968–present
Raymond Davis (United States) measures the flux of solar neutrinos and discovers that it is only a fraction (about $1/3$) of the predicted flux from expected nuclear reactions in the core of the Sun

1971
John Belcher and Leverett Davis (with earlier help from Ed Smith) (United States) demonstrate that the solar wind is full of Alfvén waves, stimulating the study of coronal heating and solar wind acceleration by waves

1972
Using data from an Orbiting Solar Observatory (OSO 4), Richard H. Munro and George L. Withbroe (United States) discover coronal holes, later shown by Werner M. Neupert and Victor Pizzo (United States) to be the source of solar wind because they correlate with magnetic storms on Earth

1972
Robert Noyes (United States) and Donald Hall (Australia) recognize that the carbon monoxide molecules on the Sun (discovered 1952) must be

colder than the gases around them, with a temperature significantly below the minimum temperature measured by other techniques

1973, June 30
John Beckman (United Kingdom) and other scientists use a Concorde supersonic passenger jet flying at 1,250 miles per hour (2,000 kilometers per hour) over Africa to extend the duration of solar eclipse totality to 74 minutes—10 times longer than can ever be observed from the ground

1973
Jack Eddy (United States) demonstrates that the sunspot cycle turned off in the 17th century, showing that the Sun changes in more ways than the 11-year sunspot cycle

1973–1974
Astronauts on NASA's Skylab orbiting laboratory study corona over a nine-month period using the Apollo Telescope Mount: (1) x-ray images show coronal holes and the relationship between these "open-field" regions with the mysterious "M-regions" postulated by Sydney Chapman and Julius Bartels (Germany) in 1940 but never seen; (2) the high-speed solar wind escaping from coronal holes produces geomagnetic disturbances when it reaches the Earth about two days after the coronal hole passes through the central meridian of the Sun; (3) the active corona is composed of loops of plasma following strong magnetic field lines, indicating that magnetism heats the corona

1973–1978
Michael Schulz (United States) begins a shift from thinking about the Sun as being magnetically sectored north–south to thinking of the sectors essentially following the solar equatorial plane. Ed Smith and coworkers confirm this discovery by using the NASA spaceprobe Pioneer 11 (1978) to observe the Sun at high solar latitudes. The magnetic sectors form a "ballerina skirt" fluttering around the Sun's equatorial plane

1979, February 26
Alan Clark (United Kingdom/Canada) and Rita Boreiko (Canada) use a NASA Learjet to observe limb occultation of the solar chromosphere in the far infrared during an eclipse

1980
NASA's Solar Maximum Mission satellite is launched; proves that the Sun varies its energy output (craft crippled by blown fuses after 9 months; repaired in orbit by Space Shuttle astronauts in 1984 and operates until 1989)

1981, 1988
Far infrared observations during eclipses from NASA's Kuiper Airborne Observatory by Eric Becklin, Charles Lindsey, and colleagues (United States) demonstrate a significant limb extension and smaller-than-predicted limb brightening of the Sun at wavelengths between 20 and 800 micrometers

1987
The first comprehensive observation of the solar infrared spectrum between 2 and 16 micrometers from space by the ATMOS Space Shuttle experiment by Crofton Farmer (United Kingdom/United States) and Robert Norton (United States)

1987

Eugene Parker (United States) proposes that the coronal magnetic field must be continuously reconnecting and releasing energy and that this mechanism can heat the corona

1988

Joseph Hollweg and Walter Johnson (United States) present a model of coronal holes in which high-frequency ion-cyclotron waves heat the corona and that the resulting fast solar wind is driven by hot protons. Although data at the time indicate that the protons are not hot, later SOHO data (John Kohl, 1995) show that the protons are hot, as predicted

1990, October

ESA/NASA Ulysses spacecraft launched from Space Shuttle into polar orbit around the Sun (via gravitational assist from Jupiter) to explore the polar regions of the Sun and its corona, which are poorly seen from Earth

1991, July 11

Astronomers observe the total eclipse from the world's largest optical observatory, Mauna Kea on Hawaii, with the 15-meter James Clerk Maxwell Telescope (United Kingdom-Netherlands-Canada), the 4-meter Canada-France-Hawaii Telescope, the 3-meter Infrared Telescope Facility (NASA), and the University of Hawaii's 2-meter telescope, making sensitive radio and infrared measurements of the solar chromosphere and obtaining high-resolution optical images of the corona and prominences

1991, August 31

Japan's Yohkoh spacecraft launched; verifies, over years of observations, that magnetic reconnection is responsible for coronal mass ejections and flares

1994, May 10

Alan Clark (United Kingdom/Canada) and coworkers use lunar limb occultation during an annular eclipse to measure profiles of spectral lines in the chromosphere formed by carbon monoxide

1995, December 2

Launch of ESA/NASA satellite SOHO (Solar and Heliospheric Observatory), leading to the discovery by Günter Brückner and colleagues that coronal mass ejections occur daily

1998, April 1

NASA TRACE (Transition Region and Coronal Explorer) satellite launched; Alan Title and Leon Golub (United States) use TRACE to provide the first high-resolution observations of the Sun from space

Glossary

annular eclipse   A central eclipse of the Sun in which the angular diameter of the Moon is too small to completely cover the disk of the Sun and a thin ring (*annulus*) of the Sun's bright apparent surface surrounds the dark disk of the Moon. Thus, an annular eclipse is actually a special kind of partial eclipse of the Sun. There are more annular eclipses than total eclipses.

annular-total eclipse   A solar eclipse that begins as an annular eclipse, changes to a total eclipse along its path, and then returns to annular before the end of the eclipse path. Also called a **hybrid eclipse**.

anomalistic month   The time it takes (27.55 days) for the Moon to orbit the Earth as measured from its closest point to Earth (perigee) to its farthest point (apogee) and back to its closest point again.

anomalistic year   The time it takes (365.26 days) for the Earth to orbit the Sun as measured from its closest point to the Sun (perihelion) to its farthest point (aphelion) and back to its closest point again.

antumbra   The extension of the Moon's shadow cone so that it forms a mirror image of itself. A region experiencing an annular eclipse lies in the antumbra of the Moon.

aphelion   The point for any object orbiting the Sun where it is farthest from the Sun.

apogee   The point for any object orbiting the Earth where it is farthest from the Earth.

arc minute   An angular measurement: 1 minute of arc is 60 seconds of arc and $1/60$ of a degree of arc.

arc second   An angular measurement: 1 second of arc is 1/60 of a minute of arc and $1/3600$ of a degree of arc.

ascending node (of the Moon)   The point on the Moon's orbit where it crosses the ecliptic (orbit of the Earth) going north.

Baily's Beads   An effect seen just before and after the total phase of a solar eclipse in which the Moon hides all the light from the Sun's disk except for a few bright points of sunlight passing through valleys at the rim of the Moon.

central eclipse (of the Sun)   An eclipse in which the axis of the Moon's shadow touches the Earth. A central eclipse can be either total or annular.

chromosphere   The reddish lower atmosphere of the Sun just above the photosphere. The chromosphere is only about 600 miles (1,000 kilometers) thick and the temperature is about 7,100°F (4,200°C).

contact (in a solar eclipse)   Special numbered stages of a solar eclipse. In a total (or annular) eclipse, *first contact* occurs when the leading edge of the Moon appears tangent to the western rim of the Sun, initiating the eclipse; *second contact* occurs when the Moon's leading edge appears tangent to the eastern rim of the Sun, initiating the total (or annular) phase of the eclipse; *third contact* occurs when the Moon's trailing edge appears tangent to the western rim of the Sun, concluding the total (or annular) phase of the eclipse; and *fourth contact* occurs when the trailing edge of the Moon appears tangent to the eastern rim of the Sun, concluding the eclipse. In a partial eclipse, there are only first and fourth contacts. These same contact points are used to describe lunar eclipses, planet transits across the face of the Sun, satellite transits across the face of a planet, and binary star transits across the face of one another.

core (of the Sun)   The central regions of the Sun where it produces its energy in a nuclear fusion reaction by converting hydrogen into helium at a temperature of 27 million°F (15 million°C).

corona   The rarefied upper atmosphere of the Sun that appears as a white halo around the totally eclipsed Sun. Speeds of atomic particles in the corona give it a temperature sometimes exceeding 2 million°F (1.1 million°C).

coronagraph   A special telescope that produces an artificial solar eclipse by masking the Sun's apparent surface with an opaque disk; invented by Bernard Lyot in 1930.

coronal hole   A region of the corona low in brightness and density. It is from coronal holes that solar particles escape most easily into space to become the solar wind.

coronal mass ejections   Vast bubbles of gas ejected from the corona into space. The bubbles can expand to a size larger than the Sun. Coronal mass ejections seem to be the principal cause of aurorae and a significant hazard to electric power grids and spacecraft electronics when these high-velocity particles and ropes of magnetic fields hit the Earth.

degree of obscuration   The fraction of the area of the Sun's disk obscured by the Moon at eclipse maximum, usually expressed as a percentage. (Degree of obscuration is not the same as the magnitude of an eclipse, which is the fraction of the Sun's diameter that is covered).

descending node (of the Moon)   The point on the Moon's orbit where it crosses the ecliptic (orbit of the Earth) going south.

Diamond Ring Effect   The stage of a solar eclipse when only one beam of light from the Sun's photosphere is reaching the Earth through a valley on the Moon.

draconic month   The time it takes (27.21 days) for the Moon to orbit the Earth as measured from ascending node through descending node and back to ascending node again. Eclipses of the Sun and Moon can take place only near a node.

eclipse limit (for the Sun)   The maximum angular distance that the Sun can be from a node of the Moon and still be involved in an eclipse seen from Earth. For partial eclipses, the limit ranges from 15°21' to 18°31' according to the varying angular sizes of the Moon and Sun due to the elliptical orbits of the Moon and Earth. For a central eclipse, the maximum and minimum limits are 11°50' and 9°55'.

eclipse maximum   The moment and position in a solar eclipse when the cone of the Moon's shadow is aimed most nearly at the center of the Earth.

eclipse season   The period of time in which the apparent motion of the Sun places it close enough to a node of the Moon so that an eclipse is possible. The Sun crosses the ascending and descending nodes of the Moon in a period of 346.62 days, so eclipse seasons occur about 173.3 days apart. Depending on where the Sun is on the ecliptic (how fast the Earth is moving), a solar eclipse season may last from 31 to 37 days.

eclipse year   The time (346.62 days) it takes for the apparent motion of the Sun to carry it from ascending node to the descending node and back to the ascending node of the Moon.

ecliptic   The apparent annual path of the Sun around the star field as seen from the Earth as the Earth orbits the Sun in the course of a year. (Thus the ecliptic is the plane of the Earth's orbit around the Sun.) The Sun's apparent path is called the ecliptic because all eclipses of the Sun and Moon occur on or very close to this track in the sky.

exeligmos   An eclipse repetition cycle of 54 years 34 days, equal to 3 saros cycles and often called the *triple saros*. After one exeligmos cycle, a solar eclipse returns to almost the same longitude but occurs about 600 miles (1,000 kilometers) north or south of its predecessor.

filament   A dark threadlike feature seen against the face of the Sun. A filament is a **prominence** seen from the top rather than at the edge of the Sun.

flare   Intense brightening in the upper atmosphere of the Sun that erupts vast amounts of charged particles into space. Flares can reach temperatures of 36 million°F (20 million °C).

greatest eclipse   The moment and position in a solar eclipse when the Moon is apparently largest in comparison to the Sun, so that the duration of the total phase of a total eclipse is *longest* and the duration of the annulus (ring of sunlight around the Moon) in an annular eclipse is *shortest*.

hybrid eclipse   An **annular-total eclipse**.

inex   A period of 10,571.95 days (29 years less 20.1 days) after which another eclipse of the Sun or Moon will occur (although not of the same type, such as total). This period equals 358 synodic months and 388.5 draconic months.

lunation   The time it takes (29.53 days) for the Moon to complete a phasing cycle (also called a *synodic period*).

magnitude (of a solar eclipse)   The fraction of the apparent diameter of the solar disk covered by the Moon at eclipse maximum. Eclipse magnitude is usually expressed as a decimal fraction: below 1.000 is a partial eclipse; 1.000 or above is a total eclipse. (The magnitude of a solar eclipse is not the same as the degree of obscuration, which is the percentage of the area of the Sun's disk that is covered.)

maximum eclipse   The moment in a solar eclipse when the shadow of the Moon passes closest to the center of the Earth. This is also the instant when the greatest fraction of the Sun's disk is obscured.

mid-eclipse   The instant in a central solar eclipse halfway between second and third contacts.

new moon   The phase of the Moon when it is most nearly in conjunction with the Sun (also called dark-of-the-moon). Solar eclipses can occur only at new moon. (In ancient times, new moon had a different meaning: the crescent Moon when it became visible after dark-of-the-moon.)

nodes   The two points at which the orbit of a celestial body crosses a reference plane. The Moon crosses the orbital plane of the Earth (**ecliptic**) going northward at the ascending node and going southward at the descending node.

partial eclipse (of the Sun)   An eclipse in which a portion of the Sun's disk is not covered by the Moon.

penumbra   The portion of a shadow from which only part of the light source is occulted by an opaque body. Seen from outside the shadow region, the penumbra is a fuzzy fringe to the dark umbra that declines in darkness outward from the umbra. A region experiencing a partial solar eclipse lies in the penumbra of the Moon.

perigee   The point for any object orbiting the Earth where it is closest to the Earth.

perihelion   The point for any object orbiting the Sun where it is closest to the Sun.

photosphere   The apparent "surface" of the Sun. It is actually a layer of hot gases only about 300 miles (500 kilometers) thick where the Sun's atmosphere changes from opaque to transparent, and visible light escapes from the Sun. The temperature of the photosphere is about 10,000°F (5,500°C).

prominence   An arch or filament of denser gas in the Sun's corona, shaped by the magnetic field of the Sun. Some prominences rise but most are descending, as if raining.

regression of the nodes   The westward shift of the Moon's nodes along the ecliptic due to tidal forces on the Moon's orbit exerted by the Sun and Earth. The regression of the nodes is responsible for the eclipse year being 18.62 days shorter than the seasonal (tropical) year. The nodes complete a westward regression entirely around the ecliptic in 18.6 years.

saros   An eclipse cycle of 6,585.32 days (18 years 11⅓ days or 18 years 10⅓ days if five leap years occur in the interval) in which an eclipse will occur that is very similar to the one that preceded it. The saros results from the near equivalence of 223 synodic months, 19 eclipse years, and 239 anomalistic months.

shadow bands   Faint flickers or ripples of light sometimes seen on the ground or buildings shortly before or after the total phase of a solar eclipse. Shadow bands are caused by light from the thin crescent of the

Sun passing through parcels of rising and falling air that have different densities and hence act as lenses to bend the light continuously in varying amounts.

solar constant   The amount of power from the Sun falling on an average square meter of the Earth's surface (1.35 kilowatts).

solar wind   A stream of charged particles (mostly protons, electrons, and helium nuclei) ejected from the Sun that flows by the Earth at 720,000 to 1.8 million miles per hour (320 to 800 kilometers per second). When enhanced by flares, the particles collide with molecules in the Earth's upper atmosphere so intensely that they cause the upper atmosphere to glow by fluorescence in displays of the aurora (the northern and southern lights).

spectrohelioscope   A solar spectroscope that blocks unwanted colors so that an observer can view the Sun in the light of one spectral line at a time; invented by Jules Janssen in 1868.

spectroscope   A device (usually employing a prism or a diffraction grating) to spread out a beam of light into its component wavelengths for study. Spectroscopy can reveal the composition, temperature, radial velocity, rotation, magnetic fields, and other features of a light source.

spicule   A jetlike spike of upward-moving gas in the chromosphere of the Sun. Viewed near the edge of the Sun, spicules resemble a forest. Each spicule lasts 10 minutes or so and ejects material into the corona at speeds of 12–19 miles per second (20–30 kilometers per second).

sunspot   A darker area in the Sun's photosphere where magnetic fields are very strong. The temperatures of sunspots are 2,500 to 3,600°F (1,400 to 2,400°C) cooler than their surroundings, making them appear darker. Sunspots can last from a day up to several months.

sunspot cycle   A period averaging 11.1 years in which the number of sunspots increases, decreases, and then begins to increase again.

synodic month   The period of time (29.53 days) required for the Moon to orbit the Earth and catch up with the Sun again. Because the Moon's position with respect to the Sun determines the phase of the Moon, the synodic period is the time required for a complete set of phases by the Moon.

transition region   The thin, irregular layer that separates the chromosphere from the corona. In this layer the temperature rises suddenly from about 7,200°F (4,000°C) to about 1.8 million °F (1 million °C). The transition region is of variable thickness, sometimes no more than tens of miles.

tritos   A period of 3,986.6295 days (11 years less 31 days) after which another eclipse will occur (although not the same type, such as total). This period, less accurate than the saros or inex, equals 135 synodic months, 146.5 draconic months, and roughly 144.5 anomalistic months.

total eclipse (of the Sun)   An eclipse in which the angular size of the Moon is sufficient to totally cover the disk of the Sun. In a total eclipse, the umbral shadow of the Moon touches the surface of the Earth.

umbra   The central, completely dark portion of a shadow from which all of the light source is occulted by an opaque body. A region experiencing a total solar eclipse lies in the umbra of the Moon.

Notes

Chapter 1: The Experience of Totality

Epigraph: Donald H. Menzel and Jay M. Pasachoff: *A Field Guide to the Stars and Planets*, 2nd edition (Boston: Houghton Mifflin, 1983), page 409.

1. In sky observations, the western side of the Sun or Moon refers to the edge of the Sun or Moon closer to the western horizon. For observers in mid-northern latitudes, the Sun is usually to the south. When facing south, east is to the left and west is to the right. This south-looking orientation can briefly confuse readers who are used to maps that are oriented north, so that east is to the right and west to the left.

2. Special thanks to John R. Beattie of New York City, upon whose experience and description this chapter is based.

Chapter 2: The Great Celestial Cover-Up

Epigraph: Alfonso X, King of Castile as cited by Arthur Koestler: *The Sleepwalkers* (New York: Grosset & Dunlap, 1963), page 69.

1. Alan D. Fiala, U.S. Naval Observatory, personal communication, April 1990.

2. The principal cause of the seasons is the tilt of the Earth's axis, not the Earth's rather modest change in distance from the Sun.

3. The Sun takes 365¼ days to appear to go through the constellations of the zodiac once around the sky, so some ancient peoples measured it or rounded it off to 360 days. Each day, the Sun goes one step around its great sky circle, which is why the Babylonians considered that a circle has 360 degrees. The Chinese considered that a circle had 365¼ degrees.

4. The use of the word *saros* to mean a 223-lunar-month eclipse cycle,

however, was erroneously introduced in 1691 by Edmond Halley when he applied it to the Babylonian eclipse cycle on the basis of a corrupt manuscript by the Roman naturalist Pliny. The Babylonian sign SAR has meaning as both a word and a number. As a word, it means (among other things) *universe*. As a number, it means 3,600, signifying a large number. But there is no evidence that the Babylonians ever applied *saros* to this 18-year eclipse cycle.

5. This period is called anomalistic because it derives from the anomaly—or irregularity—of lunar motion due to the Moon's elliptical orbit.

6. The reason for the change in latitude after 54 years 34 days is the 34-day difference from a calendar year, which changes the eclipse's position within the season, so the altitude of the Sun is significantly different and the shadow is cast farther north or south. The triple saros cycle of 54 years 34 days was known to the Chaldeans as well as to the Greeks, who called it the *exeligmos*.

7. In a sense, it has not vanished altogether. With each period of 6,585.32 days, the Sun's position with respect to the node continues to slip westward without experiencing or causing any eclipses until, after about 5,500 years, it encounters the eclipse limit of the opposite node, and that saros may be said to be reborn. George van den Bergh: *Periodicity and Variation of Solar (and Lunar) Eclipses* (Haarlem: H. D. Tjeenk Willink, 1955).

Chapter 3: A Quest to Understand

Epigraph: Jack B. Zirker: *Total Eclipses of the Sun* (New York: Van Nostrand Reinhold, 1984), page vi.

1. One of these 30 uprights is only half the diameter of all the others, as if to suggest 29½, the length of time in days that it takes the Moon to complete a cycle of phases.

2. There were also 56 chalk-filled Aubrey Holes just inside the embankment, but their function is still not generally agreed upon.

3. The hole for this stone was discovered in 1979 under the shoulder of the road that passes close to the Heel Stone. The original stone is gone. Michael W. Pitts: "Stones, Pits and Stonehenge," *Nature*, volume 290, March 5, 1981, pages 46–47.

4. The limits of the Moon's motion north and south of the ecliptic can differ by as much as 10 minutes of arc from the mean inclination of 5 degrees 8 minutes because of gravitational perturbations on the Moon caused by the Sun and the equatorial bulge of the Earth.

5. How Stonehenge may have been used as a computer to predict eclipses is proposed and explained in Gerald S. Hawkins (with John B. White): *Stonehenge Decoded* (Garden City, N.Y.: Doubleday, 1965); Gerald S. Hawkins: *Beyond Stonehenge* (New York: Harper & Row, 1973); and Fred Hoyle: *On Stonehenge* (San Francisco: W. H. Freeman, 1977).

6. James Legge, editor and translator: *The Chinese Classics*, volume 3, *The Shoo King* [Shu Ching] (Hong Kong: Hong Kong University Press, 1960), part 1, chapter 2, paragraphs 3–8 (pages 18–22). Legge romanizes Hsi's name as He. In some recountings, Hsi's name appears as Hi. The *Shu Ching* is one of five books in the *Wu Ching* (Five Classics), the sourcebooks of the Confucian tradition.

7. James Legge, editor and translator: *The Chinese Classics*, volume 3, *The Shoo King* (Hong Kong: Hong Kong University Press, 1960), part 3, book 4, chapter 2, paragraph 4 (pages 165–166; notes continue to page 168).

8. Joseph Needham and Wang Ling: *Science and Civilisation in China*, volume 3, *Mathematics and the Sciences of the Heavens and the Earth* (Cambridge: At the University Press, 1959), page 422. For the record, there were no total solar eclipses visible in China through this four-year period, and no partials of any consequence. Perhaps, if there is any historical foundation to this story, the court astronomer predicted that a major eclipse would occur and it did not, or perhaps the problem concerned lunar eclipses.

9. Joseph Needham and Wang Ling: *Science and Civilisation in China*, volume 3, *Mathematics and the Sciences of the Heavens and the Earth* (Cambridge: At the University Press, 1959), page 409. Most of this information is also available in Colin A. Ronan's abridgment of Needham's work, *The Shorter Science and Civilisation in China*, volume 2 (Cambridge: Cambridge University Press, 1981).

10. Anthony F. Aveni: *Skywatchers of Ancient Mexico* (Austin: University of Texas Press, 1980), page 181.

11. Bernardino de Sahagún: *Florentine Codex; General History of the Things of New Spain*, book 7, *The Sun, Moon, and Stars, and the Binding of the Years*, translated from the Aztec by Arthur J. O. Anderson and Charles E. Dibble (Santa Fe, New Mexico: School of American Research; Salt Lake City: University of Utah, 1953), pages 36 and 38.

Chapter 4: Eclipses in Mythology

Epigraph: Francis Baily: "Some Remarks on the Total Eclipse of the Sun, on July 8th, 1842," *Memoirs of the Royal Astronomical Society*, volume 15, 1846, page 6.

1. Viktor Stegemann: "Finsternisse," in Hanns Bächtold-Stäubli, editor: *Handwörterbuch des Deutschen Aberglaubens*, Bände 2 (Berlin: W. de Gruyter, 1930), columns 1509–1526. Germanic eclipse lore described in this chapter comes from this article, unless otherwise noted.

2. Arthur Berriedale Keith: *Indian Mythology*, volume 6 of *The Mythology of All Races* (Boston: Marshall Jones, 1917), pages 151, 192.

3. Wilhelm Max Müller: *Egyptian Mythology*, volume 12 of *The Mythology of All Races* (Boston: Marshall Jones, 1918), pages 33, 90, 124–125.

4. Paul Yves Sébillot: *Le folk-lore de France*, tome 1, *Le ciel et la terre* (Paris: Librairie orientale & américaine, 1904), page 40.

5. John C. Ferguson: *Chinese Mythology*, volume 8 of *The Mythology of All Races* (Boston: Marshall Jones, 1928), page 84. Hartley Burr Alexander: *Latin-American Mythology*, volume 11 of *The Mythology of All Races* (Boston: Marshall Jones, 1920), page 319. Mardiros H. Ananikian: *Armenian Mythology*, volume 7 of *The Mythology of All Races* (Boston: Marshall Jones, 1925), page 48.

6. Arthur Berriedale Keith: *Indian Mythology*, volume 6 of *The Mythology of All Races* (Boston: Marshall Jones, 1917), pages 232–233.

7. Mardiros H. Ananikian: *Armenian Mythology*, volume 7 of *The Mythology of All Races* (Boston: Marshall Jones, 1925), page 48.

8. Arthur Berriedale Keith: *Indian Mythology*, volume 6 of *The Mythology of All Races* (Boston: Marshall Jones, 1917), page 234.

9. Joseph Needham and Wang Ling: *Science and Civilisation in China*, volume 3, *Mathematics and the Sciences of the Heavens and the Earth* (Cambridge: At the University Press, 1959), page 228.

10. Hartley Burr Alexander: *North American Mythology*, volume 10 of *The Mythology of All Races* (Boston: Marshall Jones, 1916), pages 25, 277.

11. James George Frazer: *Balder the Beautiful*, volume 1; *The Golden Bough*, volume 10 (London: Macmillan, 1930), page 162.

12. Mabel Loomis Todd: *Total Eclipses of the Sun*, revised edition (Boston: Little, Brown, 1900), page 131.

13. Hartley Burr Alexander: *North American Mythology*, volume 10 of *The Mythology of All Races* (Boston: Marshall Jones, 1916), page 255.

14. Paul Yves Sébillot: *Le folk-lore de France*, tome 1, *Le ciel et la terre* (Paris: Librairie orientale & américaine, 1904), page 52. The chronicler was Jean Juvénal des Ursins.

15. Arthur Berriedale Keith: *Indian Mythology*, volume 6 of *The Mythology of All Races* (Boston: Marshall Jones, 1917), page 234.

16. A free translation from the "Second Soir." Compare Bernard Le Bovier de Fontenelle: *A Plurality of Worlds*, translated by John Glanvill (England: Nonesuch Press, 1929), with the excerpt in François Arago: *Popular Astronomy*, volume 2, translated by W. H. Smyth and Robert Grant (London: Longman, Brown, Green, Longmans, and Roberts, 1858), page 349.

17. Hartley Burr Alexander: *Latin-American Mythology*, volume 11 of *The Mythology of All Races* (Boston: Marshall Jones, 1920), pages 277–278.

18. James George Frazer: *The Magic Art*, volume 1; *The Golden Bough*, volume 1 (London: Macmillan, 1926), page 311.

19. Patrick Menget: "30 juin 1973: station de Surinam," *Soleil est mort; l'éclipse totale de soleil du 30 juin 1973* (Nanterre, France: Laboratoire d'ethnologie et de sociologie comparative, 1979), pages 119–142.

20. Hartley Burr Alexander: *Latin-American Mythology*, volume 11 of *The Mythology of All Races* (Boston: Marshall Jones, 1920), page 135.

21. Hartley Burr Alexander: *Latin-American Mythology*, volume 11 of *The Mythology of All Races* (Boston: Marshall Jones, 1920), page 82.

Chapter 5: Strange Behavior of Man and Beast

Epigraph: John Milton: *Paradise Lost*, book 1, lines 594 and 597–599. See John Milton: *The Complete Poems* (New York: Crown Publishers, 1936), page 24.

1. Herodotus: *The History*, volume 1, translated by George Rawlinson; Everyman's Library, volume 405 (London: J. M. Dent, 1910), book 1, chapter 74, pages 36–37.

2. Robert R. Newton lists three annular eclipses seen in the region during a 50-year period that he feels are equally likely to have given rise to this story, although an annular eclipse is not nearly as spectacular as one that is total. *Ancient Astronomical Observations and the Accelerations of the Earth and Moon* (Baltimore: Johns Hopkins Press, 1970), pages 94–97.

3. Herodotus: *The History*, volume 2, translated by George Rawlinson; Everyman's Library, volume 406 (London: J. M. Dent, 1910), book 7, chapter 37, page 136.

4. Plutarch: *The Rise and Fall of Athens: Nine Greek Lives*, translated by Ian Scott-Kilvert (Baltimore: Penguin Books, 1960), pages 201–202. Ironically,

Pericles' raid was disastrous for the Athenian forces; they fell victim to the plague. Pericles was fined and temporarily stripped of power.

5. Thucydides: *History of the Peloponnesian War*, translated by Richard Crawley (New York: E. P. Dutton, 1910), book 2, paragraph 28. Stars would not have been visible at Athens. Perhaps Thucydides heard reports from where the eclipse was annular and incorporated them into his account.

6. From the chapter "Second soir" in Bernard Le Bovier de Fontenelle: *Entretiens sur la pluralité des mondes* (Paris: Chez la veuve C. Blageart, 1686), cited in François Arago: *Popular Astronomy*, volume 2, translated by W. H. Smyth and Robert Grant (London: Longman, Brown, Green, Longmans, and Roberts, 1858), pages 359–360.

7. Mabel Loomis Todd: *Total Eclipses of the Sun*, revised edition (Boston: Little, Brown, 1900), pages 141–142.

8. F. Richard Stephenson and David H. Clark: *Applications of Early Astronomical Records*, Monographs on Astronomical Subjects, number 4 (New York: Oxford University Press, 1978), page 9.

9. F. Richard Stephenson and David H. Clark: *Applications of Early Astronomical Records*, Monographs on Astronomical Subjects, number 4 (New York: Oxford University Press, 1978), page 14.

10. François Arago: *Popular Astronomy*, volume 2, translated by W. H. Smyth and Robert Grant (London: Longman, Brown, Green, Longmans, and Roberts, 1858), page 359.

11. François Arago: *Popular Astronomy*, volume 2, translated by W. H. Smyth and Robert Grant (London: Longman, Brown, Green, Longmans, and Roberts, 1858), page 362.

12. William J. S. Lockyer: "The Total Eclipse of the Sun, April 1911, as Observed at Vavau, Tonga Islands," in Bernard Lovell, editor: *Astronomy*, volume 2, The Royal Institution Library of Science (Barking, Essex: Elsevier Publishing, 1970), pages 190–191.

Chapter 6: Anatomy of the Sun

Epigraph: Amos 8:9, *New American Standard Bible*.

1. Solar physicists Joseph Hollweg, Charles Lindsey, and Jay Pasachoff calculated or reviewed the statistics and information in this chapter. Personal communication, October 20-November 3, 1998. Lindsey calculates that 4.26 million metric tons per second (4.69 million English tons per second) of mass are converted to energy in the core of the Sun.

2. The figure of 14.8 trillion years is based on the Sun converting *all* its mass to energy, which, of course, it cannot do. Only 0.7 percent of the hydrogen mass is converted into energy; the rest becomes helium. If only the total amount of hydrogen available is considered, the Sun's life span falls to about 77 billion years. Of course, only the hydrogen close to the center of the Sun is under sufficient pressure so that the temperatures are great enough for fusion to occur. Only about 10 percent of the Sun's hydrogen will undergo fusion, reducing the Sun's lifetime to 8–10 billion years.

To generate and sustain a hydrogen-to-helium fusion reaction, the core of a star must have a temperature of at least 18 million °F (10 million °C). To have enough gravity to generate sufficient pressure to obtain this high a core temperature, a star must have at least 8 percent the mass of the Sun.

3. The expressions "fire-ocean" to describe the chromosphere and "flame-

brushes" to describe features in the corona were used by Agnes M. Clerke: *A Popular History of Astronomy during the Nineteenth Century*, 4th edition (London: A. and C. Black, 1902), pages 68, 175.

4. Spectroscopically, the red of the chromosphere is produced by the hydrogen-alpha line.

5. Solar tornadoes were discovered in 1998 by David Pike and Helen Mason using the Solar and Heliospheric Observatory (SOHO), a collaboration of the European Space Agency (ESA) and NASA (United States).

6. The temperature of the corona can be misleading. The Sun's magnetic fields cause the electrically charged atoms of the corona to move at great speeds (high temperature), but the density of these ions is so low that the corona has relatively little heat (energy in a given volume). The corona is so rarefied that if you had a box there 100 miles (160 kilometers) on each side (1 million cubic miles; 4.1 million cubic kilometers), you would entrap less than a pound (0.4 kilogram) of matter. The corona is a good vacuum by laboratory standards on Earth.

Solar physicist Joe Hollweg points out that it is also hard to explain the heating of the chromosphere, which requires as much energy as the corona. Personal communication, August 3, 1998.

7. Coronal mass ejections were discovered using NASA's Orbiting Solar Observatory 7 between 1971 and 1973 and confirmed using NASA's Solar Maximum Mission satellite in 1980 and after 1984, following repair in orbit by Space Shuttle astronauts. In the 1990s, a new generation of solar spacecraft studied corona mass ejections: Yohkoh (Japan), SOHO (*Solar and Heliospheric Observatory*—ESA and NASA), and TRACE (*Transition Region and Coronal Explorer*—NASA).

8. At sunspot minimum, about one coronal mass ejection a week is observed. Near sunspot maximum, two or three coronal mass ejections are observed each day on the average.

9. Coronal structures trace out the magnetic field lines, just as iron filings trace out the field around a bar magnet.

Chapter 7: The First Eclipse Expeditions

Epigraph: Francis Baily: "Some Remarks on the Total Eclipse of the Sun, on July 8th, 1842," *Memoirs of the Royal Astronomical Society*, volume 15, 1846, page 4.

1. Francis Baily: "On a Remarkable Phenomenon That Occurs in Total and Annular Eclipses of the Sun," *Memoirs of the Royal Astronomical Society*, volume 10, 1838, pages 1–42. Baily, searching back through the records, realized that Edmond Halley in 1715 and many other observers had seen and reported this beadlike apparition before him. Among the previous observers of the beads, he named José Joaquin de Ferrer, who also described the corona and first called it by that name. Perhaps it was Ferrer's account and his use of the word *corona* that came to Baily's mind when he saw the eclipse of 1842.

2. This and the following Baily quotations are from Francis Baily: "Some Remarks on the Total Eclipse of the Sun, on July 8th, 1842," *Memoirs of the Royal Astronomical Society*, volume 15, 1846, pages 1–8.

3. Agnes M. Clerke: "Baily, Francis," *The Dictionary of National Biography*, volume 1 (London: Oxford University Press, 1921), page 903.

4. Joseph Needham and Wang Ling: *Science and Civilisation in China*, vol-

ume 3, *Mathematics and the Sciences of the Heavens and the Earth* (Cambridge: At the University Press, 1959), page 423.

5. José Joaquin de Ferrer: "Observations of the Eclipse of the Sun, June 16th, 1806, Made at Kinderhook, in the State of New-York," *Transactions of the American Philosophical Society*, volume 6, 1809, pages 264–275.

6. Dorrit Hoffleit: *Some Firsts in Astronomical Photography* (Cambridge, Massachusetts: Harvard College Observatory, 1950). The first successful photograph of the uneclipsed Sun, also a daguerreotype, was achieved by the French scientists Hippolyte Fizeau and Léon Foucault in 1845.

7. De La Rue made that discovery in 1861. Earlier evidence for sunspots as depressions had come from observations by Scottish astronomer Alexander Wilson in 1774. He noted that the geometry of sunspots seemed to change as they were seen from different angles as the Sun's rotation carried them across the solar disk and toward the limb.

8. The Sun as a sphere of hot gas had been proposed independently by Angelo Secchi and John Frederick William Herschel (William's son) in 1864.

9. J. Norman Lockyer: "On Recent Discoveries in Solar Physics Made by Means of the Spectroscope," in Bernard Lovell, editor: *Astronomy*, volume 1, The Royal Institution Library of Science (Barking, Essex: Elsevier Publishing, 1970), page 90.

10. Alfred Fowler: "Sir Norman Lockyer, K.C.B., 1836–1920," *Proceedings of the Royal Society of London*, series A, volume 104, December 1, 1923, pages i–xiv.

11. Auguste Comte: *The Essential Comte, Selected from Cours de philosophie positive*, translated by Margaret Clarke (London: Croom Helm, 1974), pages 74, 76.

12. A. J. Meadows: *Science and Controversy: A Biography of Sir Norman Lockyer* (London: Macmillan, 1972), page 53.

13. J. Norman Lockyer: "On Recent Discoveries in Solar Physics Made by Means of the Spectroscope," in Bernard Lovell, editor: *Astronomy*, volume 1, The Royal Institution Library of Science (Barking, Essex: Elsevier Publishing, 1970), pages 101–102.

14. "A gravely mutilated state" is the picturesque description found in Gabrielle Camille Flammarion and André Danjon, editors: *The Flammarion Book of Astronomy*, translated by Annabel and Bernard Pagel (New York: Simon and Schuster, 1964), page 227.

Chapter 8: The Eclipse That Made Einstein Famous

Epigraph: Arthur S. Eddington as cited in Allie Vibert Douglas: *The Life of Arthur Stanley Eddington* (London: T. Nelson, 1956), page 44.

1. The angular displacement of a star is inversely proportional to the angular distance of that star from the Sun's center.

2. In 1911, Einstein had asked Freundlich to investigate another aspect of his emerging general theory of relativity: the motion of Mercury. Freundlich reviewed the anomaly in Mercury's motion, recognized since the work of Le Verrier in 1859, and reconfirmed that Mercury was indeed deviating slightly from the law of gravity as formulated by Newton. Freundlich's results, which matched Einstein's relativistic recalculation for the precession of Mercury's orbit, were published in 1913 over the objections of his superiors.

3. To make certain that the telescope optics and the camera system introduced no unknown deflection in the positions of stars, it was important to have for comparison with the eclipse plate a picture of that same region taken with the same equipment when the Sun was not present. No such plates were available.

4. Charles Dillon Perrine, an American who was directing the Argentine National Observatory, prepared to test the bending of starlight during the October 10, 1912, eclipse in Brazil, but was rained out.

5. Ronald W. Clark: *Einstein, the Life and Times* (London: Hodder and Stroughton, 1973), page 176.

6. Freundlich's team was not alone in the Crimea. An American team from the Lick Observatory also journeyed to Russia to test relativity, but they were clouded out. In 1918, Freundlich left the Royal Observatory to work full-time with Einstein at the Kaiser Wilhelm Institute. In 1920, he was appointed observer and then chief observer and professor of astrophysics at the newly created Einstein Institute at the Astrophysical Observatory, Potsdam. Freundlich continued to be plagued by miserable luck on his eclipse expeditions to measure the deflection of starlight. He returned empty-handed in 1922 and 1926 because of bad weather. He finally got to see an eclipse in Sumatra in 1929, although he obtained a deflection (now known to be erroneous) considerably greater than Einstein predicted. When Hitler came to power, Freundlich left Germany and eventually settled in Scotland, where he changed his name to Finlay-Freundlich, based on his mother's maiden name, Finlayson. He built the first Schmidt-Cassegrain telescope, the prototype for almost all large photographic survey telescopes today.

7. Banesh Hoffmann with the collaboration of Helen Dukas: *Albert Einstein, Creator and Rebel* (New York: Viking Press, 1972), page 116.

8. Banesh Hoffmann with the collaboration of Helen Dukas: *Albert Einstein, Creator and Rebel* (New York: Viking Press, 1972), page 125.

9. Robert V. Pound and Glen A. Rebka, Jr.: "Resonant Absorption of the 14.4-kev Gamma Ray from 0.10-microsecond Fe^{57}," *Physical Review Letters*, volume 3, December 15, 1959, pages 554–556. The gravitational redshift was detected previously, based on Einstein's suggestion, in light emitted from extremely dense white dwarf stars, but to nowhere near the same degree of certainty.

10. The Lick team was working with mediocre equipment and improvised mounts. Their excellent regular equipment had stood ready under cloudy skies in Russia in 1914, but it was too cumbersome to transport home after the outbreak of World War I. After the war, it was delayed in shipment home and missed the eclipse of 1918.

11. The "Newtonian prediction" is calculated on the basis that light energy has a mass equivalent (Einstein's $E = mc^2$). That mass is then treated as ordinary matter using Newton's equations for gravity.

12. Allie Vibert Douglas: *The Life of Arthur Stanley Eddington* (London: T. Nelson, 1956), pages 40–41.

13. Ronald W. Clark: *Einstein, the Life and Times* (London: Hodder and Stroughton, 1973), page 226.

14. Frank W. Dyson, Andrew C. D. Crommelin, and Arthur S. Eddington: "Joint Eclipse Meeting of the Royal Society and the Royal Astronomical Soci-

ety," *The Observatory*, volume 42, November 1919, pages 389–398. The article includes dissenting comments from scientists in audience. The article based on eclipse results was Frank W. Dyson, Arthur S. Eddington, and Charles R. Davidson: "A Determination of the Deflection of Light by the Sun's Gravitational Field, From Observations Made at the Total Eclipse of May 29, 1919," *Philosophical Transactions of the Royal Society of London*, series A, volume 220, 1920, pages 291–333. A summary of the eclipse results had appeared the week after the joint meeting: Andrew C. D. Crommelin: "Results of the Total Solar Eclipse of May 29 and the Relativity Theory," *Nature*, volume 104, November 13, 1919, pages 280–281. Eddington had yet another reason to be proud: "By standing foremost in testing, and ultimately verifying, the 'enemy' theory, our national observatory kept alive the finest traditions of science; and the lesson is perhaps still needed in the world today" (Clark: *Einstein*, page 284). British physicist Robert Lawson made a similar observation: "The fact that a theory formulated by a German has been confirmed by observations on the part of Englishmen has brought the possibility of cooperation between these two scientifically minded nations much closer. Quite apart from the great scientific value of his brilliant theory, Einstein has done mankind an incalculable service" (Clark: *Einstein*, page 297). The world soon forgot this lesson.

15. Jeffrey Crelinsten: "William Wallace Campbell and the 'Einstein Problem': An Observational Astronomer Confronts the Theory of Relativity," *Historical Studies in the Physical Sciences*, volume 14, part 1, 1983, pages 1–91. Clarence Augustus Chant of Canada also measured a deflection favoring Einstein's theory.

16. Allie Vibert Douglas: *The Life of Arthur Stanley Eddington* (London: T. Nelson, 1956), page 44.

17. Jack B. Zirker: *Total Eclipses of the Sun* (New York: Van Nostrand Reinhold, 1984), pages 177–178.

18. Edward B. Fomalont and Richard A. Sramek: "A Confirmation of Einstein's General Theory of Relativity by Measuring the Bending of Microwave Radiation in the Gravitational Field of the Sun," *Astrophysical Journal*, volume 199, August 1, 1975, pages 749–755. The result reported in their article, 1.775 ± 0.019 arc seconds, was later revised to 1.761 ± 0.016 arc seconds. (Personal communication, March 1990.) They used three 85-foot (26-meter) antennas and a distant 45-foot (14-meter) antenna, instruments of the National Radio Astronomy Observatory at Green Bank, West Virginia. The idea of using radio waves and radio telescopes to test the bending of light was first proposed by Irwin I. Shapiro: "New Method for the Detection of Light Deflection by Solar Gravity," *Science*, volume 157, August 18, 1967, pages 806–807.

Chapter 9: Modern Scientific Uses for Eclipses

Epigraph: Robert Louis Stevenson: "My Shadow," lines 1–2, in *A Child's Garden of Verses* (New York: C. Scribner's Sons, 1905), page 23.

1. Sabatino Sofia, David W. Dunham, Joan B. Dunham, and Alan D. Fiala: "Solar Radius Change Between 1925 and 1979," *Nature*, volume 304, August 11, 1983, pages 522–526.

2. Some changes observed on the Sun may have been reflected in modest climate changes on Earth within historical times. From 1645 to 1715, there were very few sunspots. During this period, called the Maunder minimum, temperatures in Europe were lower than usual, leading some modern scientists to refer to these seven decades as the Little Ice Age.

3. He gathers data from edge observations of annular eclipses as well.

4. The Moon appears to keep the same side always facing the Earth, because it spins once around on its axis as it revolves once around our planet. But the Moon's orbital speed is not constant, because its orbit is elliptical. The Moon moves faster when closer to the Earth and slower when farther away. As a result, the Moon does not quite keep one side precisely aligned with Earth. This slight misalignment provides different horizons along the eastern and western edges of the Moon.

The 5° inclination of the Moon's orbit to the ecliptic means that as the Moon approaches the Sun for an eclipse, it is ascending or descending through the Sun-Earth plane. Because observers on Earth see the Moon crossing the Sun slightly top or bottom first, this perspective foreshortens or extends the view across the Moon's polar regions and changes which features cause Baily's Beads.

A lesser libration effect is the view from different locations on Earth. Depending on our latitude, we see a little over or under the Moon's poles. From the east or west, we see the polar features at slightly different angles. This change of perspective with each eclipse and each position along the eclipse path causes the apparent positions and angular heights of features to change minutely.

In addition, the Moon actually wobbles slightly due to the tidal effects of the Earth and Sun, adding in a complex way to this libration effect.

5. Jay M. Pasachoff and Brant O. Nelson: "Time of the 1984 Total Solar Eclipse and the Size of the Sun," *Solar Physics*, volume 108, 1987, pages 191–194.

6. Paraphrased from the cartoon by Rodney deSarro in Gurney Williams, editor: *I Meet Such People!* (New York: Farrar, Straus, 1946), page 20.

7. C. Lindsey, E. E. Becklin, F. Q. Orrall, M. W. Werner, J. T. Jefferies, and I. Gatley: "Extreme Limb Profiles of the Sun at Far-Infrared and Submillimeter Wavelengths," *Astrophysical Journal*, volume 308, September 1, 1986, pages 448–458; and T. L. Roellig, E. E. Becklin, J. T. Jefferies, G. A. Kopp, C. A. Lindsey, F. Q. Orrall, and M. W. Werner: "Submillimeter Solar Limb Profiles Determined from Observations of the Total Solar Eclipse of 1988 March 18," *Astrophysical Journal*, volume 381, November 1, 1991, pages 288–294; and C. Lindsey, J. T. Jefferies, T. A. Clark, R. A. Harrison, M. Carter, G. Watt, E. E. Becklin, T. L. Roellig, D. C. Braun, D. A. Naylor, and G. J. Tompkins: "Extreme-Infrared Brightness Profile of the Solar Chromosphere Obtained During the Total Eclipse of 1991," *Nature*, volume 358, July 23, 1992, pages 308–310. See also Peter Foukal: "Darkness Can Illuminate," *Nature*, volume 358, July 23, 1992, pages 285–286.

8. T. A. Clark, C. A. Lindsey, D. M. Rubin, and W. C. Livingston: "Eclipse Measurements of the Distribution of CO Emission Above the Solar Limb," in J. R. Kuhn and M. J. Penn: *Infrared Tools for Solar Astrophysics, What's Next?*,

Proceedings of the National Solar Observatory/Sacramento Peak Observatory Workshop, September 1994, pages 133–138; also Alan Clark, personal communication, October 20 and 24, 1998.

9. This collaboration included scientists from the Max Planck Institute in Germany, Rhodes College, the National Solar Observatory, Michigan State University, and the High Altitude Observatory.

10. J. R. Kuhn, M. J. Penn, and I. Mann: "The Near-Infrared Coronal Spectrum," *Astrophysical Journal*, volume 456, January 1, 1996, pages L67–L70; J. R. Kuhn, H. Lin, P. Lamy, S. Koutchmy, and R. N. Smartt: "IR Observations of the K and F Corona During the 1991 Eclipse," in D. M. Rabin et al., editors: *Infrared Solar Physics* (Netherlands: International Astronomical Union, 1994), pages 185–197; P. Lamy, J. R. Kuhn, H. Lin, S. Koutchmy, and R. N. Smartt: "No Evidence of a Circumsolar Dust Ring from Infrared Observations of the 1991 Solar Eclipse," *Science*, volume 257, September 4, 1992, pages 1377–1380.

11. Laurence J. November and Serge Koutchmy: "White-Light Coronal Dark Threads and Density Fine Structure," *Astrophysical Journal*, volume 466, July 20, 1996, pages 512–528; Serge Koutchmy and Michaël M. Molodensky: "Three-dimensional image of the solar corona from white-light observations of the 1991 eclipse," *Nature*, volume 360, December 24/31, 1992, pages 717–719.

12. S. Koutchmy, O. Bouchard, S. Grib, L. November, J-C. Vial, P. Gouttebrose, V. Koutvitsky, M. Molodensky, L. Solov'iev, and I. Veselovsky: "About Small Plasmoids Propagating in the Solar Corona," *Proceedings of the Third SOHO Workshop—Solar Dynamic Phenomena and Solar Wind Consequences* (September 26–29, 1994) (European Space Agency publication SP-373, December 1994), pages 139–142.

13. Report to the 1998 Keck Northeast Astronomy Consortium Student Research Symposium by Kevin D. Russell, with special reference to work by Tim McConnochie, supplied by Jay M. Pasachoff, October 16, 1998.

Pasachoff and Benjamin D. Knowles also used observations made by Pasachoff and Stefan Martin during the 1998 eclipse on Aruba to help Naval Research Laboratory scientists calibrate the only coronagraph aboard the ESA/NASA Solar and Heliospheric Observatory (SOHO) spacecraft capable of studying the corona close to the Sun. In a coordinated effort, they took photographs of the Sun at the same time as SOHO and in the same wavelength of light. By comparing the images, SOHO scientists could gauge how much stray light was scattering around in their coronagraph's detector and correct for that false signal in subsequent observations. Report to the 1998 Keck Northeast Astronomy Consortium Student Research Symposium by Benjamin D. Knowles, supplied by Jay M. Pasachoff, October 16, 1998.

Chapter 10: Observing a Total Eclipse

Epigraph: Rebecca R. Joslin: *Chasing Eclipses: The Total Solar Eclipses of 1905, 1914, 1925* (Boston: Walton Advertising and Printing, 1929), pages 1–2.

1. François Arago: *Popular Astronomy*, volume 2, translated by W. H. Smyth and Robert Grant (London: Longman, Brown, Green, Longmans, and Roberts, 1858), page 360.

2. Hollweg says he uses a "poor man's version" of a patch. "I stick a clean hanky between my eye and my glasses."

3. Anton Pannekoek: *A History of Astronomy* (London: G. Allen & Unwin, 1961), page 406.

4. George B. Airy: "On the Total Solar Eclipse of 1851, July 28," page 1 in Bernard Lovell, editor: *Astronomy*, volume 1, The Royal Institution Library of Science (Barking, Essex: Elsevier Publishing, 1970).

5. George B. Airy: "On the Total Solar Eclipse of 1851, July 28," page 4 in Bernard Lovell, editor: *Astronomy*, volume 1, The Royal Institution Library of Science (Barking, Essex: Elsevier Publishing, 1970). This speech was given on May 2, 1851, prior to the total eclipse on July 28.

6. Mabel Loomis Todd: *Total Eclipses of the Sun*, revised edition (Boston: Little, Brown, 1900), page 21.

7. Isabel Martin Lewis: *A Handbook of Solar Eclipses* (New York: Duffield, 1924), page 62.

8. The Moon rotates once on its axis while it revolves once around the Earth. The result is that we always see the same face of the Moon; the other half is always turned away from us. Almost, but not quite. Instead of seeing only 50 percent of the Moon, over time we actually see 59 percent due to librations. In this case, longitudinal and diurnal librations are at work: (1) The speed of the Moon's spin is constant but the speed of the Moon in its elliptical orbit varies, which allows us to see a little beyond the eastern and western limbs of the Moon in the course of a month; (2) when the Moon is near the horizon, we are looking at it a little from the side, so we are peeking a bit beyond the eastern or western limb seen when the Moon is high in the sky.

9. François Arago: *Popular Astronomy*, volume 2, translated by W. H. Smyth and Robert Grant (London: Longman, Brown, Green, Longmans, and Roberts, 1858), pages 360–361.

10. Mabel Loomis Todd: *Total Eclipses of the Sun*, revised edition (Boston: Little, Brown, 1900), page 19.

11. Rebecca R. Joslin: *Chasing Eclipses: The Total Solar Eclipses of 1905, 1914, 1925* (Boston: Walton Advertising & Printing, 1929), pages 14–15. "M" was Mary Emma Byrd, first director of the Smith College Observatory.

12. Mabel Loomis Todd: *Total Eclipses of the Sun*, revised edition (Boston: Little, Brown, 1900), page 25.

13. Mabel Loomis Todd: *Total Eclipses of the Sun*, revised edition (Boston: Little, Brown, 1900), page 174.

14. Mabel Loomis Todd: *Total Eclipses of the Sun*, revised edition (Boston: Little, Brown, 1900), page 152.

15. Associated Press, March 2, 1998, reporting on the eclipse of February 26.

16. Camille Flammarion: *The Flammarion Book of Astronomy*, edited by Gabrielle Camille Flammarion and André Danjon, translated by Annabel and Bernard Pagel (New York: Simon and Schuster, 1964), page 147.

Chapter 11: Observing Safely

Epigraph: Richard Le Gallienne: "For Sundials," cited in John Bartlett: editor: *Familiar Quotations*, 13th edition (Boston: Little, Brown, 1955), page 830.

1. Isabel Martin Lewis: *A Handbook of Solar Eclipses* (New York: Duffield, 1924), page 98.

Chapter 12: Eclipse Photography

Epigraph: Stephen J. Edberg, personal communication, March 1990.

1. Black-and-white film is useful mainly for scientific purposes.

2. Negatives can be made into slides and slides can be made into prints, but some quality may be lost in this conversion. It is best to select slide or print film according to what you plan to do with your photography.

3. For the latest information on film products, check photography magazines such as *Popular Photography*, *Outdoor Photography*, and *Petersen's Photo-Graphic*. In addition, film manufacturers and good photo stores provide technical bulletins on the latest films. You can also get film information on the manufacturers' Web sites (www.agfa.com, www.fujifilm.com, www.kodak.com, and www.konica.com).

4. For more information on safe solar filters, see B. Ralph Chou: "Solar Filter Safety," *Sky & Telescope*, volume 95, February 1998, pages 36–40.

5. The viewfinder of a single-lens reflex camera uses a prism and mirror system that allows you to look directly through the same lens that takes the actual photograph. When you press the shutter button, the mirror flips up and the shutter curtain opens, permitting the image to reach the film. An instant later, the shutter closes and the mirror returns to its original position.

6. Although zoom lenses usually have smaller apertures and are slower than fixed lenses, they work fine for photographing eclipses if they are of good optical quality.

7. Michael Covington: *Astrophotography for the Amateur* (Cambridge: Cambridge University Press, 1988), page 59.

8. Catadioptric "mirror lenses" are made by all the major camera manufacturers (Canon, Minolta, Nikon, Pentax, etc.), as well as independent lens makers (Phoenix, Sigma, Tamron, Tokina, Vivitar, etc.).

9. Manufacturers of apochromatic, fluorite, and extra-low-dispersion refractors include AstroPhysics, Meade, Takahashi, Tele-Vue, and Vixen.

10. If your clock drive requires AC, check on a source of electricity at your observing site or use a drive corrector with a built-in converter that changes 12-volt DC to 60-cycle, 110-volt AC. Many countries use 220-volt AC electricity. If you want to plug into their power, bring along an AC converter, available in many hardware stores.

11. This quick-and-dirty polar alignment should be good enough for the eclipse. If you need a better alignment, you might try a technique called drift polar alignment (sunearth.gsfc.nasa.gov/eclipse/align.html).

12. Michael Covington: *Astrophotography for the Amateur* (Cambridge: Cambridge University Press, 1988), page 95.

13. Bogen/Manfrotto, Gitzo, and Slik all make heavy-duty tripods.

14. Some photographers like to expose one extra stop during the thin crescent phases because the Sun's limb is a little darker than disk center. This step is important only if you shoot slide film because it has a smaller exposure latitude.

15. For film ISOs or f-numbers not listed, use the following formula to determine your exposure.

$E = f^2/(A \times B)$

E = exposure or shutter speed (seconds)
f = focal ratio
A = film speed (ISO)
B = brightness value (see table below)

Brightness Value (approximate)	Eclipse Feature
256	Partial and annular phases (using solar filter)
128	Prominences and innermost corona
32–128	Diamond Ring Effect
32	Inner corona
4	Middle corona
0.5	Outer corona

For example, taking a picture of prominences on the Sun during totality ($B = 100$) using a telephoto lens at f/14 with 400 ISO film would require an exposure of:

$E = 14^2/(400 \times 100)$
$E = 196/40,000$
$E = 0.0049$ second or 1/204 second

Normal SLRs don't have an exposure of 1/204 second, so use the closest shutter speeds, which would be 1/125 and 1/250 second.

16. Here's a strategy used by many eclipse photographers to plan their exposures. Using the film ISO and telescope or lens f-number, determine the shortest shutter speed to capture the prominences (relatively bright) and longest shutter speed for capturing the outer corona (dim). After totality begins, shoot a sequence of exposures using every shutter speed, starting with the one for prominences and ending with the one for the outer corona. For instance, for ISO 400 and f/5.6, the recommended exposure for prominences is 1/1000 and for the outer corona 1/8 second. The shutter-speed sequence would then be: 1/1000, 1/500, 1/250, 1/125, 1/60, 1/30, 1/15, and 1/8. This is a total of eight exposures. If you want some additional insurance, you could now repeat the sequence in reverse, ending with 1/1000. With a 36-exposure roll of film, you would still have 20 exposures for the diamond ring and prominences at the beginning and end of totality.

The exposures given here have been determined after photographing more than a dozen solar eclipses, but they are only suggestions. Each eclipse is different and the corona's brightness varies. Weather conditions (haze or thin clouds) may require longer exposure times. Use the recommended exposures as a starting point and then bracket during the eclipse, especially if the weather is a factor.

17. The price of GPS receivers has dropped dramatically during the 1990s. They are now available for under $100.

18. George B. Airy: "On the Total Solar Eclipse of 1851, July 28," *Astron-*

omy, volume 1, Bernard Lovell, editor: The Royal Institution Library of Science (Barking, Essex: Elsevier Publishing, 1970), page 4.

19. See especially Fred Espenak's eclipse Web site (sunearth.gsfc.nasa.gov /eclipse/eclipse.html).

20. Contact the International Occultation Timing Association (www.anomalies.com/iota/splash.htm) for details. See also chapter 9 in this book.

Chapter 13: The Pedigree of an Eclipse

Epigraph: George B. Airy: "On the Total Solar Eclipse of 1851, July 28," page 1 in Bernard Lovell, editor: *Astronomy*, volume 1, The Royal Institution Library of Science (Barking, Essex: Elsevier Publishing, 1970).

Chapter 14: The Eclipse of August 11, 1999

Epigraph: Isabel Martin Lewis: *A Handbook of Solar Eclipses* (New York: Duffield, 1924), page 3.

1. At present, the leading (but not the only) theory of the Moon's formation involves a collision between a large planetesimal and the Earth.

2. In the case of the total eclipse of August 11, 1999, the axis of the shadow cone passes 2,006 miles (3,227 kilometers) north of the center of the Earth.

3. Estimated by the authors based on the population located on and close to the path of totality.

Selected
Bibliography

Allen, David; and Allen, Carol. *Eclipse*. Sydney; Boston: Allen & Unwin, 1987.

Arago, François. *Popular Astronomy*. 2 volumes. Translated by W. H. Smyth and Robert Grant. London: Longman, Brown, Green, Longmans, and Roberts, 1858.

Ashbrook, Joseph. *The Astronomical Scrapbook*. Edited by Leif J. Robinson. Cambridge: Cambridge University Press; Cambridge, Massachusetts: Sky Publishing, 1984.

Astrophotography Basics. (Kodak Publication P-150.) Rochester, N.Y.: Eastman Kodak, 1988.

Aveni, Anthony F. *Skywatchers of Ancient Mexico*. Austin: University of Texas Press, 1980.

Baily, Francis. "On a Remarkable Phenomenon that Occurs in Total and Annular Eclipses of the Sun." *Memoirs of the Royal Astronomical Society*, volume 10, 1838, pages 1–40.

Baily, Francis. "Some Remarks on the Total Eclipse of the Sun, on July 8th, 1842." *Memoirs of the Royal Astronomical Society*, volume 15, 1846, pages 1–8.

Brewer, Bryan. *Eclipse*. Seattle: Earth View, 1978; 2nd edition 1991.

Chambers, George F. *The Story of Eclipses*. Library of Valuable Knowledge. New York: D. Appleton, 1912.

Clerke, Agnes M. *A Popular History of Astronomy During the Nineteenth Century*. 4th edition. London: A. and C. Black, 1902.

Couderc, Paul. *Les éclipses*. (Que sais-je? series no. 940.) Paris: Presses Universitaires de France, 1961.

Covington, Michael. *Astrophotography for the Amateur*. Cambridge: Cambridge University Press, 1988.

Douglas, Allie Vibert. *The Life of Arthur Stanley Eddington*. London: T. Nelson, 1956.

Dyson, Frank; and Richard v.d. R. Woolley. *Eclipses of the Sun and Moon*. Oxford: At the Clarendon Press, 1937.

Espenak, Fred. *Fifty Year Canon of Solar Eclipses: 1986–2035*. Washington, D.C.: NASA; Cambridge, Massachusetts: Sky Publishing, 1987. NASA Reference Publication 1178 Revised.

Espenak, Fred. "Predictions for the Total Solar Eclipse of 1991," *Journal of the Royal Astronomical Society of Canada*, volume 83, June 1989, pages 157–178.

Espenak, Fred; and Jay Anderson. *Total Solar Eclipse of 1999 August 11*. NASA Reference Publication 1398, Washington, D.C.: NASA, 1997.

Fiala, Alan D.; James A. DeYoung; and Marie R. Lukac. *Solar Eclipses, 1991–2000*. (U.S. Naval Observatory circular 170.) Washington, D.C.: U.S. Naval Observatory, 1986.

Flammarion, Camille. *The Flammarion Book of Astronomy*. Edited by Gabrielle Camille Flammarion and André Danjon. Translated by Annabel and Bernard Pagel. New York: Simon and Schuster, 1964.

Francillon, Gérard; and Patrick Menget, editors. *Soleil est mort: l'éclipse totale de soleil du 30 Juin 1973*. Nanterre: Laboratoire d'ethnologie et de sociologie comparative (Récherches thématiques, 1), 1979.

Friedman, Herbert. *Sun and Earth*. New York: W. H. Freeman (Scientific American Library), 1986.

Golub, Leon; and Jay M. Pasachoff. *The Solar Corona*. Cambridge: Cambridge University Press, 1997.

Guillermier, Pierre; and Serge Koutchmy. *Éclipses totales: Histoire, Découvertes, Observations*. Paris: Masson, 1998.

Harrington, Philip S. *Eclipse! The What, Where, When, Why, and How Guide to Watching Solar and Lunar Eclipses*. New York: John Wiley & Sons, 1997.

Harris, Joel; and Richard Talcott. *Chasing the Shadow*. Waukesha, Wisconsin: Kalmbach, 1994.

Johnson, Samuel J. *Eclipses, Past and Future; with General Hints for Observing the Heavens*. Oxford: J. Parker, 1874.

Joslin, Rebecca R. *Chasing Eclipses: The Total Solar Eclipses of 1905, 1914, 1925*. Boston: Walton Advertising and Printing, 1929.

Kippenhahn, Rudolph. *Discovering the Secrets of the Sun*. New York: John Wiley & Sons, 1994.

Kudlek, Manfred; and Erich H. Mickler. *Solar and Lunar Eclipses of the Ancient Near East from 3000 B.C. to 0 with Maps*. Neukirchen-Vluyn, Germany: Butzon & Bercker Kevelaer, 1971.

Lang, Kenneth R. *Sun, Earth, and Sky*. New York: Springer, 1995.

Lang, Kenneth R.; and Owen Gingerich, editors. *A Source Book in Astronomy and Astrophysics, 1900–1975*. Cambridge, Massachusetts: Harvard University Press, 1979.

Lewis, Isabel M. *A Handbook of Solar Eclipses*. New York: Duffield, 1924.

Lewis, Isabel M. "The Maximum Duration of a Total Solar Eclipse." *Publications of the American Astronomical Society*, volume 6, 1931, pages 265–266.

Little, Robert T. *Astrophotography: A Step-by-Step Approach*. New York: Macmillan, 1986.

Lovell, Bernard, editor. *Astronomy*, 2 volumes. The Royal Institution Library of Science. Barking, Essex; New York: Elsevier Publishing, 1970.

Marschall, Laurence A. "A Tale of Two Eclipses." *Sky & Telescope*, volume 57, February 1979, pages 116–118.

Maunder, Michael. "Eclipse Chasing" (on eclipse photography), pages 139–157 in Patrick Moore, editor. *Yearbook of Astronomy*. New York: W. W. Norton, 1989.

Meadows, A. J. *Early Solar Physics*. Oxford: Pergamon Press, 1970.

Meeus, Jean. *Astronomical Algorithms*. Richmond, Virginia: Willmann-Bell, 1991.

Meeus, Jean. *Elements of Solar Eclipses: 1951–2200*. Richmond, Virginia: Willmann-Bell, 1989.

Meeus, Jean. *Mathematical Astronomy Morsels*. Richmond, Virginia: Willmann-Bell, 1997.

Meeus, Jean; Carl C. Grosjean; and Willy Vanderleen. *Canon of Solar Eclipses* (solar eclipses from 1898 to 2510 A.D.). Oxford: Pergamon Press, 1966.

Menzel, Donald H.; and Jay M. Pasachoff. *A Field Guide to the Stars and Planets*, 2nd edition. Boston: Houghton Mifflin, 1983.

Mitchell, Samuel A. *Eclipses of the Sun*, 5th edition. New York: Columbia University Press, 1951.

Mucke, Hermann; and Jean Meeus. *Canon of Solar Eclipses—2003 to 2526*. Vienna: Astronomisches Büro, 1983.

Needham, Joseph; and Wang Ling. *Science and Civilisation in China*. Volume 3: *Mathematics and the Sciences of the Heavens and the Earth*. Cambridge: At the University Press, 1959.

Neugebauer, Otto. *The Exact Sciences in Antiquity*, 2nd edition. Providence: Brown University Press, 1957.

Newton, Robert R. *Ancient Astronomical Observations and the Accelerations of the Earth and Moon*. Baltimore: Johns Hopkins Press, 1970.

Oppolzer, Theodor von. *Canon of Eclipses*, (solar and lunar eclipses from 1207 B.C. to 2161 A.D.). Translated by Owen Gingerich. New York: Dover, 1962.

Osterbrock, Donald E.; John R. Gustafson; and W. J. Shiloh Unruh. *Eye on the Sky: Lick Observatory's First Century*. Berkeley: University of California Press, 1988.

Ottewell, Guy. *Astronomical Calendar*. Greenville, North Carolina: Astronomical Workshop, annually.

Ottewell, Guy. *The Understanding of Eclipses*. Greenville, North Carolina: Astronomical Workshop, 1991.

Pasachoff, Jay M., Michael A. Covington. *The Cambridge Eclipse Photography Guide*. Cambridge: Cambridge University Press, 1993.

Pepin, R. O.; J. A. Eddy; and R. B. Merrill, editors. *The Ancient Sun: Fossil Record in the Earth, Moon and Meteorites*. Proceedings of the Conference on the Ancient Sun; Boulder, Colorado; October 16–19, 1979. New York: Pergamon Press, 1980.

Rao, Joe. *Your Guide to the Great Solar Eclipse of 1991*. Cambridge, Massachusetts: Sky Publishing, 1989.

Reynolds, Michael D.; and Richard A. Sweetsir. *Observe: Eclipses*. Washington, D.C.: Astronomical League, 1995.

Sébillot, Paul Y. *Le folk-lore de France*. Volume 1: *Le ciel et la terre*. Paris: Librairie orientale & américaine, 1904.

Silverman, Sam; and Gary Mullen. "Eclipses: A Literature of Misadventures." *Natural History*, volume 81, June-July 1972, pages 48–51, 82.

Stegemann, Viktor. "Finsternisse," in Hanns Bächtold-Stäubli, editor. *Handwörterbuch des Deutschen Aberglaubens*, volume 2. Berlin: W. de Gruyter, 1930, columns 1509–1526.

Stephenson, F. Richard; and David H. Clark. *Applications of Early Astronomical Records*. Monographs on Astronomical Subjects, 4. New York: Oxford University Press, 1978.

Thompson, J. Eric S. *A Commentary on the Dresden Codex; A Maya Hieroglyphic Book*. Philadelphia: American Philosophical Society, 1972.

Todd, Mabel Loomis. *Total Eclipses of the Sun*. Revised. Boston: Little, Brown, 1900.

Wentzel, Donat G. *The Restless Sun*. Washington, D.C.: Smithsonian Institution Press, 1989.

Zirker, Jack B. *Total Eclipses of the Sun*, 2nd edition. Princeton: Princeton University Press, 1995.

Index